WHERE THE *light* BEGINS

WHERE THE *Light* BEGINS

Stories of Near-Death Experiences and After-Death Communications by Children and How They Experience Them Differently Than Adults

Daniela Klose and Ralph Klose

The information provided in this book is designed to provide helpful information on the subjects discussed. This book is not meant to be used, nor should it be used, to diagnose or treat any medical condition. The author and publisher are not responsible for any specific health needs that may require medical supervision and are not liable for any damages or negative consequences from any treatment, action, application, or preparation, to any person reading or following the information in this book.

References are provided for information purposes only and do not constitute endorsement of any websites or other sources. In the event you use any of the information in this book for yourself, the author and the publisher assume no responsibility for your actions.

Books may be purchased through booksellers or by contacting Sacred Stories Publishing.

Where the Light Begins
Stories of Near-Death Experiences and After-Death Communications by Children and How They Experience Them Differently Than Adults

Daniela Klose and Ralph Klose

Tradepaper Print ISBN: 978-1-958921-91-3
Electronic Book ISBN: 978-1-958921-92-0
Library of Congress Control Number: 2026937001

Published by Sacred Stories Publishing

Printed in the United States of America

This book is for the children who returned from the threshold of death and carry memories of a greater light within them.

TABLE OF CONTENTS

How Children Experience Consciousness Differently

Sometimes children say things that defy explanation. Remarkable observations of what happens when consciousness loosens its hold on the body. They speak of a light that is not brightness, but presence; and describe moments in which their familiar sense of self dissolves while something else quietly emerges.

The accounts might begin quietly. A child says over breakfast:

"I think I came back into something that no longer fits me."

A gaze lingers as if reaching beyond time. They draw circles of light or a bridge that leads nowhere and everywhere. And when they recount what they have experienced, their language is sensory, intuitive, and image based—yet the words carry emotional coherence and extraordinary depth.

Near-death experiences have long challenged the boundaries of medicine, psychology, and neuroscience. They occur under extreme conditions, such as cardiac arrest, trauma, or terminal illness, when we might expect consciousness to be absent. And yet, the reports from those who survived—detailing radiant light, panoramic memory, or lucid perception—are consistent across cultures. Their descriptions of these experiences open a window into a deeper layer of life, one that transcends the traditional boundaries between self and other, between body and spirit.

Children speak differently about death. They do not describe it as a boundary, an end, or an enemy but as a space, a meeting, or a return. They often speak of it unprompted, without fear, and with a clarity that unsettles and illuminates. In some cases, the child's experience does not come with a clear narrative but reflects a presence—or more precisely, a resonance, a tone, perception. They might describe not only what they saw or remembered, but also how the experience has shaped the way they interact with the world.

Children's words are also quieter, deeper than any argument. Unlike adults, they often speak without religious, metaphysical, or doctrinal framing. Their descriptions are direct; yet they are symbolic, sometimes beyond language, and notably constant. Children's words do not seek persuasion. They simply describe what they experience, without context or expectation. They might not have a memory of what was learned, but a sense of something returned to, the kind of knowledge that comes before language can shape their ideas.

This kind of remembering moves in a dimension where rational thought falls silent. When we truly listen to them, something rare opens—a light of stillness where concepts give way to contact, and a new understanding of consciousness begins.

In transpersonal psychology, such experiences are sometimes described as *pre-symbolic*, felt and known before we have words, concepts, or explanations for them. In children, this mode of awareness is especially visible.

Young children have not yet fully developed the cognitive structures responsible for linear thought, self-reflection, and moral evaluation. They might not yet experience a sense of self in the way adults do. Perhaps this openness, the absence of internal filters, allows more directness. What in adults is overlaid by fear, narrative, or belief appears in children in its most essential form. Sound becomes color. Light becomes presence. Time softens. The boundary between inside and outside dissolves. And what remains is the kind of knowledge that cannot be taught: "I was there" or "And that was good."

What Science Tells Us

Near-death experiences occur regardless of a person's religious background, including in people who are unfamiliar with terms such as heaven, angel, or life after death. The children's statements are not to be understood as formulations borrowed from others, but rather as emerging from within themselves. For this very reason, because they are not shaped by belief systems, children's near-death experiences are considered particularly insightful in research. They arise from direct, immediate experience.

This directness is not naïve. It is unfiltered. It reveals a dimension of experience that resists classification because it requires none. What children describe is a form of knowledge and a memory of something that lies deeper than words. They point to a mode of perception that exists beyond neural patterns or cultural conditioning, a form of awareness that they encounter.

Classical science has offered us many answers about children's thought processes, through measurements, electrical impulses, neuronal activity patterns, and chemical signatures in the blood. And yet one question remains, one that no instrument can fully answer: What do children truly experience? What about the inner world that cannot be captured in a diagram?

Experience rarely follows a formula. A child says: "It was light, but it wasn't bright" or "It was quiet but not empty." And sometimes, these sentences appear: "I was no longer me" or "Everything was there, all at once."

How can we explain that? Perhaps not through neurochemistry, or through synaptic networks. Perhaps only through something deeper, which we might call resonance.

Many children also describe impressions that feel like a remembering, but of things they could never have lived, such as colors, spaces, voices, patterns, or a unique kind of light. Perhaps memory is not a function of neural storage, rather access to a deeper pattern that is remembered because it was never lost. When children say those sentences, they are not confused. They speak from a zone beyond language.

What if what we call the self does not reside in the brain, but in a nonlocal field, a conscious structure beyond space that manifests through the brain but is not produced by it? When the self dissolves and yet everything is felt, we might enter a realm where particles have no boundaries, and the light is a remembering from within.

Current neuroscience locates consciousness in neural networks, electrical patterns, and biochemical processes. Under general anesthesia or during cardiac arrest, when brain activity falls below the threshold of integration, conventional models predict that consciousness ceases completely. However, the studies by critical-care physician and near-death researcher Dr. Sam Parnia on cardiac arrest survivors show that vivid and structured inner experiences do occur, even in the absence of measurable cortical activity.

Researchers such as Melvin Morse, Pim van Lommel, and Bruce Greyson have documented cases where children reported details they could not have seen or heard, such as descriptions of deceased relatives they had never known or scenes they had not witnessed. Some researchers interpret these experiences as neurological artifacts such as oxygen deprivation, brain chemistry shifts, or psychological defense mechanisms. Others suggest that such experiences challenge the assumption that consciousness arises solely from brain function. The idea of nonlocal consciousness, which researchers in transpersonal psychology, quantum theory, and philosophy of mind explore, opens a space of wonder.

What if consciousness is a current rather than a possession? We are neither its origin nor its goal, but rather, a shining point in its flow. Carl Jung spoke of the *collective unconscious* as a shared psychic realm where archetypal images, symbols, patterns, and human forms reside that are not learned but remembered. What if this collective field is psychologically and physically real? It would be a consciousness space in which we are all connected, through pattern shapes of being.

Perhaps we sense this field most clearly when matter falls silent, because the brain stops filtering, death approaches, or life momentarily pauses. Perhaps what we call "near-death" was never an end but a brief glimpse into the order behind all things.

How These Stories Reached Us

The accounts that follow came to us gradually, over years, through conversations, drawings, and recollections. Time and again, we were struck by the clarity of the simplest sentences and statements. Some of the most vivid reports emerged from our long-standing work as palliative care counselors in hospices. We heard these stories in quiet moments with families, caregivers, and sometimes the children themselves. Based on our experience in the fields of psychology, cognitive science, and consciousness research, we began to ask ourselves a question: What underlying causes could give rise to these accounts?

Children have described near-death experiences, after-death communications, and the precognition of future events across cultural, familial, and medical contexts. *Where the Light Begins* addresses all three. Some of these accounts stem from structured interviews and research while others are from personal accounts or therapeutic conversations. All names and identifying details have been changed to preserve privacy. What remains is true. Each case is accompanied by a scientific commentary that helps contextualize what the child has experienced and explores possible causes. The reflection that follows each thematic set of cases invites contemplation.

Our work grounded in psychology, medicine, the humanities, and personal engagement is not seeking proof, and this book does not present a theory. It is a movement toward experiences that resist language and yet call to be shared. The cases discussed in this book invite us to rethink how we talk about life, ourselves, and reality. We do not need to name everything to recognize it. Sometimes it is enough to pause and listen to our inner voice. What if the child within us was never lost but just silent? What we know most deeply might not be knowledge but a light that sustains, recognizes, and provides orientation: a movement, a breath, a glance at the sky, a calmness within us.

We approach these stories with an openness to hear what calls from within. We are listening to what these experiences suggest about consciousness, identity, and time.

It is now an invitation to you: to pause, to listen, and to remember.

Family Members, Nature, and Animals

When children recount their near-death experiences, their words are vivid and clear, yet not analytical. They are lived moments. They do not explain; they touch. And perhaps because of this, they move us more deeply than explanation ever could. Their power lies in what they awaken within us: an echo, a resonance, and a quiet recognition that we never fully lost.

Perhaps children do not recall specific events but an in-between, a state without names. A presence beyond time and space.

In many accounts of near-death experiences, individuals report profound encounters with nature: endless skies, flowing rivers, or radiant, all-encompassing sunlight. In such accounts, nature does not merely serve as a background but as a vivid and meaningful force.

Likewise, many people describe encounters with other beings during these threshold experiences. These beings range from hard-to-describe entities to deceased loved ones who appear as comforting companions, often bearing reassuring messages or simply offering a silent, yet deeply felt, presence.

Animals also have their place in some near-death experiences. Beloved pets that have passed away make an appearance, or animals emerge in symbolic forms: snakes, bears, lions, or horses such as those described in the accounts of indigenous cultures, where they play a significant spiritual role.

In accounts of adults' near-death experiences, animals tend to be just one part of a broader scenario, often together with deceased relatives or other beings. But when children relate their near-death experiences, animals or deceased pets often appear as central attachment figures and emotional companions in the transition from life to death.

Animals and pets can also play a significant role after a child returns from the threshold of death. Some children who came back from this brink form a connection with animals that goes beyond the usual child-animal bond.

Eli

The Boy and the Girl in the Light

At the age of ten, Eli's life came to an abrupt halt. An accident happened on his way home from school. It was a winter day with wet roads and poor visibility, and a driver failed to see the child on his bicycle.

In a single moment, everything changed. Eli sustained severe internal injuries, a concussion, and a brief cardiac arrest. He was resuscitated at the scene. In the hospital, doctors placed him in a medically induced coma to stabilize his condition and to protect his brain. For thirty-nine hours, he remained unconscious. During the state of unconsciousness, something happened that no one could explain.

When Eli awoke, he was calm and quiet. He was not disoriented or crying. At first, he did not speak. Later, when he began to talk, his words were simple, clear, and undramatic.

"I was in a bright place," he said. "There was a girl and she knew me. And I knew right away she was my sister."

He described how the girl smiled at him and held his hand. There were no words—just presence and light.

"I had the feeling," he later added, "that she was there to tell me everything was okay."

For his parents, the story was unsettling. Eli was considered an only child. He had never been told of a sister. However, that night while he lay unconscious in the hospital, his half-sister—a child from one of his father's earlier relationships—died in a car accident in another city. Eli had never met her. He had never heard her name. Her photo was not present in the home. The news of her death reached the family only after Eli had already been admitted to the hospital.

Later, when he was cautiously shown a photograph, he responded without hesitation.

"That's her!" he said. "That's how she looked when she held my hand."

He never spoke much about the experience, and he did not ask questions. However, he seemed more attentive than he had been prior to his accident. He listened with greater calmness, and he responded with fewer words and deeper clarity. He noticed things more quickly. When asked whether he was afraid of death, he paused and replied with quiet certainty:

"No. Not anymore. It doesn't feel like an end. It feels like meeting again."

Scientific Commentary

Eli's experience belongs to the rare but strikingly documented cases of so-called "Peak-in-Darien" reports, near-death experiences in which persons see deceased individuals whose death was unknown to them at the time of the experience. These cases are considered particularly difficult to explain as they appear to access information that would not normally be available to consciousness under ordinary circumstances.

From a clinical perspective, Eli was unconscious due to a traumatic brain injury, internal bleeding, and a medically induced coma. Such states are considered devoid of perception. Cortical activity is reduced and sensory stimuli barely register. However, Eli reports—in a clear, specific, and emotional way—an encounter with a deceased person whose existence was previously unknown to him. He recognized the girl as his sister, who coincidentally died at the very same time. He knew who she was without any connection, prior knowledge, or external indication.

From the perspective of conventional neuroscience, such a case presents a serious challenge. There are no neurological mechanisms to explain the reported content, nor any known biochemical or electrical processes that would allow a transmission of information of this kind across spatial distance and without any sensory input. Even episodic memory has no explanatory basis here. What remains is the possibility of nonlocal perception, a form of consciousness which is not based on neural storage, but on connection.

In quantum physics, there are models that—at least theoretically—allow such phenomena. Concepts such as *entanglement*, *nonlocality*, and *morphic fields* point to a reality in which information is shared simultaneously. Biologist Rupert Sheldrake describes these fields as *patterns of resonance*: structures that remain connected across time and space, while being capable of influencing one another. In this sense, consciousness might not arise in isolation within the brain but be part of a larger field with which the brain sustains relationship.

Eli did not say much, and he did not interpret or seek explanations. His words were precise, shaped by certainty. This is exactly what makes children's near-death experiences so unique: They are calm. They do not interpret—they reveal. They do not describe ideas of an afterlife but instead testify to an experience.

Tiago

The River That Knew My Name

This transcultural account from the Brazilian rainforest concerns Tiago, who was eight years old when a severe infection brought him to the edge of life. Tiago lived with his family in a small, remote village near the Amazon River. Community life followed the rhythm of that river. The days rose and fell with the movement of water. The nights were shaped by the voices of insects, owls, and unseen things in the jungle. When Tiago became ill, no one spoke of fever. They said: "His body is resting." His breathing grew weaker and his skin turned cool.

During the night, his mother placed her hand on his forehead and began to sing. The melody was not formal or learned, and it was transmitted more through rhythm than words. Later, she said that in that moment, "He was no longer with us." Days later, Tiago spoke of his experience with a voice that sounded not as if he had invented something, but as if he remembered it.

"I wasn't in the house," he said. "I was in the river."

He described how the water carried him. The soft water was not cold or fast, as if it knew how to hold him. There was light that did not come from the sun or the sky. It shone from the current itself.

"It glowed from below," Tiago said. "And it knew my name. It called me, not with voice, but with waves."

Then an animal came: a large, silent jaguar. Its eyes were luminous, not in a threatening way but in one that conveyed awareness.

"He knew that I knew him," Tiago said. "And I wasn't surprised. I was just quiet."

Together, they moved through a jungle that was not the same as before. The trees seemed to breathe, and their leaves murmured names. The air had a pulse. Everywhere, life was not separate but connected, expressing itself through scent, warmth, or mere presence. At the edge of a clearing stood an old woman without a face. Tiago knew that she was his grandmother. However, it was not the one he had known in life.

"She smelled like fire that no longer burns," he said.

She did not speak. She simply placed her hand on his chest, and at that moment, Tiago heard the river again. However this time, it no longer called his name but his mother's. The call was not spoken aloud but conveyed through vibration and memory in the water. Alongside the call, the light returned.

"It was as if I had brought something back that belonged to her," he said.

Tiago awoke a few hours later. He was still weak and pale but calm. His breath was slow and even. His words were few, but he smiled more frequently. When his mother asked if he had been afraid, he looked at her and simply said:

"No. I was with everything that lives."

Scientific Commentary

What children in Western accounts describe as light, beings, or spaces appears in Tiago's case as a river, an animal, and an old woman without a face. However, the essence is not diminished. It is only remembered or embodied differently through resonance. In many indigenous traditions, what Western psychology calls "consciousness" is not an isolated mind-state. It lives in the breath of the wind. It pulses in the animal. It rests in the tree trunk. A child who leaves their body briefly does not step out of life but enters a deeper form of it. The boundary between the living and the transcendent beings is circular.

From a psychological perspective, these experiences may be described through the concept of *archetypes*, deeply rooted forms of awareness that appear across cultures and epochs. The jaguar, the grandmother figure, the river that speaks are expressions of what Carl Jung called the *collective unconscious*, shared symbolic structures that emerge where language ends and perception deepens. They appear not to inform but to reorient—not as metaphor, but as lived structure.

In developmental psychology, such experiences are sometimes referred to as *preverbal memory*: experiences stored before the acquisition of language. These memories are accessed through sensation, inner pictures, or rhythm. Many spiritual traditions recognize this kind of memory as well, describing it as "seeing without eyes" or "knowing without words." It is a different kind of realism that was never formed by language and therefore it is not bound by it.

Toma

The Call of the Jaguar

Toma was ten years old when he was stung by a venomous insect, deep in the Amazon rainforest. The reaction was immediate and violent. His body swelled and collapsed. His family, part of an indigenous community, carried him for hours to a remote health post, as they were far from the nearest hospital. There, Toma lost consciousness, and for more than ten minutes, he showed no sign of life.

Fortunately, he returned, and with him came the description of an experience that left not only his family, but also the elders of his people, speechless. Toma spoke neither of light nor of a tunnel. He told of a path through an unusual forest. The trees were enormous, ancient, and they were breathing. Their trunks pulsed and their leaves whispered. He heard voices within himself, without using his ears. These voices spoke in rhythms and meanings instead of words.

Then an animal, a calm and powerful black jaguar, emerged from the undergrowth. The jaguar looked at Toma. In that gaze was everything: question, answer, love, and trial. Toma followed him. They moved through the shifting light between the trees.

"I wasn't walking beside him," Toma said later. "I was inside him. And he was inside me."

At the end of the path was a circle of light who sang. Toma understood intuitively that he could either stay or return.

"The jaguar didn't ask. But I knew what he meant."

He thought of his sister and of the song she had taught him. He also thought of his mother. In that moment, the wind changed, and something carried him back.

When Toma awoke, he was different. He did not speak at first, but when he finally told what he had seen, the elders understood that this had been a return from the edge of death.

He did not seek to convince anyone of what had happened, and he did not retell his experience with urgency. Sometimes, he says things such as:

"When the jaguar calls you, he doesn't want you to come. He wants you to remember" or "The jaguar is not something you see. He is something that sees you. And if you remember that gaze, you'll never walk alone again."

Since then, Toma watches animals differently, and he hears the breath of trees. He speaks with animals as if language were unnecessary. He places his hand on the soil as if listening. When other people ask him what he learned, he often shakes his head.

"I didn't learn anything," he says. "I just stopped forgetting."

Scientific Commentary

Toma's experience touches a domain of consciousness that lies beyond verbal interpretation. It is a space in which identity feels permeable. The merging with the jaguar, the act of walking "within him," and the sense of being seen by a being who is also himself, all point to a form of ego dissolution, as described in mystical traditions and contemplative states of consciousness. The neurological interpretation of this might be a temporary suspension of the ego boundary in the prefrontal cortex. However, such an explanation is insufficient if we are to comprehend the experience in a coherent context.

Many indigenous cultures view the jaguar as a living presence that guards thresholds and conveys insight through encounter. In Toma's case, the jaguar does not appear as a projection but as a co-knowing companion, a figure that does not instruct but reminds. This form of inner communication beyond language—through gaze, movement, and atmosphere—is recognized in modern psychology as a genuine expression of nonverbal intelligence.

From an anthropological perspective, Toma's report bears a strong resemblance to initiatory experiences, namely transitional rites in which children undergo symbolically charged threshold experiences that open them to new levels of awareness. The elders' response, in the form of their silence and their understanding without explanation, suggests that they saw Toma's near-death experience as a reconnection to a deeper, collective knowing.

Such experiences challenge conventional Western explanatory models. They are nonlinear, objectively unmeasurable, and empirically irreplicable. However, they carry an inner coherence that becomes tangible through authenticity.

Emil

The Boy the Animals Found

It was a summer day when eight-year-old Emil almost drowned in a lake. There was no storm or turbulent water. In a single moment, just one pause in breath too long beneath the surface, and suddenly, he was gone. No one noticed what was happening right away. When they found him, his body was motionless and showed no pulse and no response. They attempted resuscitation. A rescue helicopter came.

Emil survived, but something had changed.

"It's like I hear what isn't being said," he said later, a quiet description of a state that required no explanation.

In the weeks that followed, his world shifted. Emil avoided large spaces, crowds, and harsh light. The sounds of the city tired him. Words felt too loud and movements felt too fast. He sought out trees, animals, and soil. His hands rested often in the grass, his forehead against the bark of a tree. His parents noticed it first with the family dog. Emil would sit beside him for hours, not speaking, just breathing. The dog would rest its head on his lap, as if Emil had become a second heartbeat. There were no commands and no games, just quiet presence.

Then other animals came. Cats approached him, unbidden. Birds landed beside him on benches. A wounded deer stood in the garden and did not flee. Emil only spoke about it once.

While sitting barefoot in soft moss, he whispered, "They know I was there. Where they come from, too."

His teacher wrote in his report: "Emil seems to feel what others are feeling before they know it themselves." Sometimes in class, he would drift away, gazing out the window as if he were listening, not to voices, but to something beneath them. He sensed a kind of "inner weather" more clearly than the world around him.

He spoke of his near-death experience only in fragments, describing a place that smelled of earth but glowed from within, of light that did not blind but warmed, of sounds like roots moving through time.

"It wasn't Heaven," he said. "It was like a forest made of sound."

And there were no people there, only stillness and life. He did not describe animals as forms or images, but as a presence that was not separate or nameable. It was just a feeling: connection without boundary, a kind of living awareness that needed no language.

After his near-death experience, Emil's drawings changed. He did not paint scenes. He drew fields, energetic spaces, and lines that touched instead of separating. He also drew vibrations in the form of waves flowing into each other, like a stag with points of light in its antlers or a beetle surrounded by a golden ring. He drew a child with bare feet and roots growing from the toes, listening to the ground. His pictures had no perspective; nevertheless, they had depth and meaning. He never said he understood animals and never claimed to speak to them. He simply said:

"They know I was there."

Once, after a long summer day in the grass, he whispered:

"They don't ask questions. They remember differently—not with thoughts, but with their whole being. When they look at me, it's not because they want something. It's because they've seen that place too, the one where everything breathes together."

His mother once found him sitting by a fox that had wandered to the edge of the woods. The boy and the animal did not touch or move. They just watched each other—quiet, still, and present. When she asked him afterward what had happened, Emil only said of the fox:

"I think it knew I wasn't entirely back. Just mostly."

Since that day, he sometimes places his hand on tree trunks and closes his eyes, seeming to listen. Those who see him rarely interrupt, because something in his stillness feels like a door that has once been open and might open again, but only in silence.

Scientific Commentary

Children who return from the light sometimes come back with an increased awareness of nature. They react more intensely to it. This is expressed in hugging trees, sensing the presence of plants, or finding a deeper sense of calm in the forest. The children become more sensitive or intuitive with animals. They feel more understood by them and often experience a kind of nonverbal communication.

After a near-death experience, the nervous system is often overactive. According to the Polyvagal Theory by psychologist and neuroscientist Stephen W. Porges, contact with elements of nature can support the regulation of the nervous system. As a result, the heart rate decreases and the parasympathetic vagus nerve, which runs from the brain down into the abdominal area, is activated. The fight-flight-freeze response is reduced.

Especially in moments of stress, grief, or emotional strain, people report certain animals approaching them more frequently.

Although animals do not possess the same linguistic capacity for verbal expression as humans, studies show that many species are able to recognize, interpret, and respond to emotional signals, which is a form of resonance perception between humans and animals. This state, which occurs independently of language, is called *embodied communication.*

Mammals possess extremely sensitive sensory systems and can perceive microscopic changes in a person's behavior and physiology, as their perceptual limits for acoustic, olfactory, and visual changes are significantly lower. In humans under stress, hormones such as cortisol alter body odor. Breathing rates increase subtly, and the heartbeat accelerates.

Studies show that interactions between humans and animals increase oxytocin levels, a hormone associated with bonding and feelings of safety, in both species. This hormone plays a key role in triggering caregiving behavior, trust, and calming responses. Animals that have formed a stable bond with a human react to that person's stress or sadness in a way similar to young children. They seek proximity, offer physical contact, and try to help restore emotional and neuroendocrine balance. Wild animals, including foxes, can distinguish between neutral people and those who are dangerous. In some cases, they even recognize individual humans. Feral pigeons have also displayed this behavior.

Resonance theories in consciousness research, such as the General Resonance Theory, propose that oscillations are present in all forms of matter. Neuronal activity patterns can emerge within specific frequency ranges. Essential aspects of conscious experience, particularly in mammals, are linked to different forms of neuronal synchronization. When different units oscillate at the same frequency, their waves interact and amplify one another, leading to increased flows of energy and information. These electrical activity patterns generate an electromagnetic field that may serve as the locus of sensory experience.

REFLECTIONS

At thresholds, empirical science begins to reach its limits and may point, tentatively, toward philosophy. When children return from a near-death experience, they often do not recount. They pause. They draw. They fall silent. And then, sometimes, a single image emerges, such as a light, a river, or a sound. Perhaps memory is not only what has happened, but also what awaits us.

Perhaps images, sounds, animals, and silence or subtle perception are not vague impressions, but part of a language we have never forgotten, only covered. And perhaps something ancient stirs within us when we truly listen to children, in order to receive. A recognition. A pause. A return to what we once knew before we learned to explain.

Eli's parents found their son's experience almost incomprehensible. An unknown half-sister died at the same time, and Eli's response was a quiet, simple recognition. No doubt and no speculation. Just a sentence: "That's her." In that sentence lay not only recognition but also acknowledgment, perhaps even of something greater than what we are able to know.

Tiago's story opens a different reality, one in which no human voice is heard and yet everything carries meaning. Tiago heard his name not as a call but as a current. The river glows from within. An animal appears, not as a threat, not as a symbol, but as a living presence. A name is not spoken but recognized. The world around him does not instruct; rather, it responds, it listens, and it remembers. For Toma, the jaguar was not a guide but a view, a presence, and a knowing. He remembered through calmness. In this way, Toma's experience does not end in his return. It continues through his steps, his presence, his listening. For him, the jaguar is no longer an image. It is a breath. A silent companion that walks behind all things. Not watching but waiting to be remembered again.

Such experiences are not religious in the narrow sense. They are not doctrinal. They follow no ideology, and yet they carry a depth that is close to the sacred. They are spiritual in the oldest meaning of the English word *spirit*, which is based on the Latin *spiritus*, breath and mind. They show that what is most essential does not belong to a system. It belongs to being itself. And what is whispered by a child in the Brazilian rainforest is not so different from what another child says in a European hospital room—with closed eyes, with an open heart.

These experiences go beyond cultural boundaries. They challenge the notion that there is a Western or indigenous version of transcendence. There are only languages of experience. Some are shaped by water, wind, and trees. Others by light, quietness, or voices. However, what resonates within them is strikingly similar: the feeling of connectedness, of recognition, of homecoming. In reports from children around the world, whether urban or rural, religious or secular, the same motifs appear: beings, animals, light, river, gaze. Some have names, others do not. They all carry the same imprint: "I was held. I was known. I was meant."

In Emil's world, silence does not merely exist. It breathes. It pulses in every moment of stillness, in every quiet glance he shares with the animals, Earth, and the wind. There is a presence that is gentle, waiting to be recognized. Emil no longer tells his story. Yet sometimes, when the wind shifts and the dog lifts its head, he listens again—not with his ears but with something primeval. A memory of silence. A sound without tone. And somewhere beneath his feet, the world breathes back. Some people call it intuition, others, sensitivity. Perhaps it is simply another kind of "ear." It is not an ear in the head but in the skin, in the chest, and in the breath. An ear that does not listen to words but to a sensed aliveness.

It is possible that within this silence lies a deeper rhythm, one we have forgotten how to listen to. Vibrations that bind everything together, threads that weave us into the world. This silence is not empty; it is full of life. It speaks through being present, through a knowing that requires no explanation. Children like Emil show us that the quietest moments hold the greatest truths. They remind us that life does not always speak in explanations but in the spaces between words, in the breath of the wind, in the gaze of an animal. In the simple act of being, in listening to the silent pulse of existence, we are most alive.

And perhaps animals also know this sound. The sound of someone who has crossed a threshold and returned quieter. A dog that lies down next to a child who almost died. A bird lands on the windowsill, not for food but because something is right. Not every encounter needs a reason; some are simply recognition. Animals do not need to believe. They feel. When they encounter a child like Emil, they move closer due to a deep attunement. When Emil says, "They know I was there," he does not mean understanding. He means connection through closeness and shared silence.

Language as a Space of Memory

Sometimes the experience of being at the threshold of death changes a basic part of the child's identity: their relationship with words. What vanishes during these inward journeys is not the body but the sense of a distinct self.

We think in language. We remember in language. We dream, pray, and sing within it. And yet its origin remains a mystery. Linguistics has devised models, family trees, and phonetic laws, but the origin of language remains a mystery. What if the proto-language did not emerge from thought but from listening? What if it was created not to describe but to connect?

Language is more than a means of communication. It is our access to the world. What happens when a child, after returning, no longer experiences the world as description but as vibration? When a sentence no longer explains anything but becomes unnecessary? What happens when a child comes back and speaks words they have never learned? When language is no longer form but meaning or pure remembrance?

Many children who return after a near-death experience do not speak as they did before. They speak differently, less, or not at all. And sometimes they express sentences that are openings. These children bring back a way of listening that does not end. An attentiveness that does not impose itself and a language that sometimes begins by becoming silent.

Liva

The Words She Had Never Learned

Liva had to undergo a complex, high-risk cranial surgical procedure when she was just ten years old. The anesthesia lasted nearly five hours. Twice, her vital signs became unstable, leading to brief collapses. Nevertheless, the doctors spoke of a "good return."

Liva came back—but she was different. She was calm and alert. Then she began to speak in words that no one understood, fluent sentences in a language unknown to her. She spoke with ease, effortlessly, and in sounds neither her parents nor her teachers had ever heard. It sounded like a blend of Old Italian and liturgical Hebrew. Some tones seemed Asian, others whispered like wind through trees, soft, floating, and rhythmic. And yet it had structure and made sense.

A nurse with Romanian roots happened to be nearby. He paused. "She's saying: I passed through the gate. It was made of breath. I've come back."

And then: "Everything is sound. Even what you think."

Later, in conversation with a linguist, no definitive classification of Liva's new language could be made. The syllables were structured and grammatically coherent but not assignable to any known language.

And still, they sounded like memory, like something older than experience and deeper than knowledge.

Liva herself said little about what she had gone through. Only this: "I didn't learn it. I knew it."

And: "It was like a language without a land but with depth."

Over time, the strange words faded. They spoke themselves out of her like an echo from within, gradually growing quieter. Even so, they remained as a trace. And sometimes, when she felt sad, she would whisper them again, softly and barely audible. She whispered them not as a message but as a tone of home.

Scientific Commentary

Liva's experience leads into a domain that neither neurolinguistics nor classical psychology can fully explain: the sudden appearance of unfamiliar but structurally coherent language following an altered state of consciousness. In scientific literature, such phenomena are rarely documented. Terms such as *xenoglossy*, the ability to speak a language never learned, or *glossolalia*, ecstatic, usually meaningless speech, do not capture her experience.

From a neurobiological perspective, one might point to linguist Noam Chomsky's theory of *generative grammar*, the idea that humans possess a universal language module. In extreme states, unconscious patterns of this syntax might be activated, resulting in sound fragments without a lexicon, yet rhythmic and internally coherent.

Cognitive science also knows of phenomena in which trance-like conditions produce language-like sequences. Yet Liva's words did not seem random. They were poetic, and they sounded like messages, carried by a symbolism that went beyond the individual.

Maybe she entered a field of memory, a space of archetypal sounds, akin to Carl Jung's concept of the collective unconscious as described earlier. Some spiritual traditions say that the universe itself is sound: a primordial creative tone, *Nada Brahma*, the world as vibration. Maybe it was precisely this that Liva experienced: a moment in which language revealed itself not as word but as affective synchrony. Even in the womb, children do respond to rhythm, pitch, and vibration. Language begins in listening as a subtle, non-verbal synchronization of emotional states.

Philosopher Maurice Merleau-Ponty described language as a pre-personal and pre-cultural form of expression, an audible unfolding of consciousness and not merely a cultural instrument. In altered states of consciousness, this unfolding can detach itself from the culturally learned form and emerge from a pre-linguistic layer of awareness.

Spiritual traditions describe concepts that point to a primordial or non-human form of sound, word, or light. In Judaism, the concept of *dabar*, the Divine word, is understood as the creative language of God. Especially in Kabbalah, the mystical tradition of Judaism, there is the idea of the "light of the letters" or the "primordial language of being." In Hinduism, there is the belief that sound is the origin of the world and of consciousness. *Śabda Brahman* refers to "the Divine sound as the highest reality."

In Buddhism as well, the primordial sound is considered an expression of non-conceptual consciousness. It can be anchored by using a mantra, which is a repeated syllable, word, phrase, or short sentence. Many indigenous cultures also understand sound, tones, rhythmic patterns, or songs not as human interventions, but as "encounters with a deeper order of the world." These include, for example, Australian songlines, North American vision songs, and Amazonian *icaros*.

Mark

The Child Who Understood the Voices

Eleven-year-old Mark was brought to the emergency room in shock due to a severe allergic reaction. He had to be resuscitated and for eight minutes he was clinically unconscious. The electroencephalogram (EEG) showed flat lines. His breathing was supported by a machine. The situation was critical, but not hopeless. Mark survived. His heartbeat returned. His breath came back into his body.

However, something in him had changed. When he began to speak again, everyone noticed it instantly. His voice was calm, alert, and bright, but the words he first spoke sounded foreign. It was a single sentence in a language that neither his parents nor the nurses understood.

Days later, Mark's speech would be identified as an archaic form of Old French, a language Mark had never studied nor heard. In the days that followed, he spoke "normally" again. And yet, stray sentences appeared, particularly brief phrases in Greek, Sanskrit, liturgical Hebrew, or something that sounded like a blend of all three. Not everything was identifiable, but some of it felt like memory. Mark did not seem surprised, puzzled, or proud. He simply said:

"I didn't know they were languages. I was just answering."

When asked what he had experienced, he shook his head.

"I wasn't in the body. I was—open. And then there was something. It didn't speak. But it was voice."

He spoke of a kind of communication that used no words and yet meant everything.

"It was like thoughts that weren't mine but meant for me. I understood them, not because I knew the language, but because they were meant for me."

He said it felt "like music you don't hear but know." He described a woven fabric of meaning with no sentences and no sound, but full understanding.

Since his return, something had shifted in him. He became quieter, not withdrawn, but more composed. His speech became unusually precise. He began writing poems, not classical in style but in images, symbols, and echoes. He once said:

"Thinking has a different language. And I heard it once."

On a visit to an old church, he suddenly stopped, closed his eyes, and whispered:

"They are praying. But not to God. To a memory."

Scientific Commentary

Mark's experience joins a small but significant group of documented cases where children do not return with visions, but with linguistic patterns and symbols that are foreign to their biography, yet inwardly familiar.

This is the phenomenon, mentioned previously, called *xenoglossy*, the spontaneous speaking of languages never learned. Yet in children like Mark, a different phenomenon emerges. It does not resemble a performance or suggestion; rather, it reflects a quiet authenticity. Language does not appear as a tool but as a field of resonance. As

Mark himself put it: "I didn't know they were languages. I was just answering."

This statement points to a dimension in which language is not used; instead, it passes through the human being. Meaning does not arise from grammar but from attunement to another level of consciousness. It is a form of language that does not operate as a learned code but as something that has always been present, a resonance that becomes audible once the self steps aside. Such experiences can hardly be explained through biography or education. They break open the models of cognitive psychology.

In transpersonal psychology, thinkers such as psychiatrist Stanislav Grof and philosopher Jean Gebser describe language as expressing deeper layers of consciousness, not merely cognitively constructed. Depth psychology also offers points of approach. Carl Jung's concept of the collective unconscious, as discussed earlier, encompasses primordial patterns of language. An archetypal language that is not learned but remembered. From this perspective, Mark's experience is an expression of a collective field of consciousness: His medical emergency was a moment in which a child became part of a language that "already knows everything."

Neuroscientific findings show that in extraordinary states of consciousness, the *default mode network*, which is responsible for the narrative construction of the self, is downregulated, while brain areas for symbolic integration and emotional resonance become more active. In particular, parietal networks, temporal association areas, and limbic and paralimbic regions show increased activity, all of which are involved in symbol processing, meaning attribution, imagistic thinking, and archetypal perceptions. This may explain why children like Mark do not hear words but find themselves within a stream of meaning, comparable to a piece of music not heard but known.

Quantum-informed theories, such as those proposed by physicist Amit Goswami or consciousness researchers Stuart Hameroff and Roger Penrose, also suggest that language does not merely arise in the brain but emerges as resonance within a deeper field. In this view, words appear not as an invention but as a fragment of memory. Goswami assumes that consciousness is primary. Matter, brain, thought, and language arise from consciousness, not the other way around. According to him, language is not a neural product; this indicates that meanings exist prior to their formulation as possibilities within the quantum world. Human beings access these potentials through intuition and creativity.

In their theory of Orchestrated Objective Reduction, Hameroff and Penrose propose that consciousness arises from quantum-coupled oscillations in the microtubules of nerve cells. However, these oscillations include components that are nonlocally entangled and may be connected to information beyond their immediate location, namely with a deeper information field, the so-called "Platonic realm of mathematical truth." Insights, meanings, and creative formulations arise when the brain resonates with this field.

Roya

The Girl Who Spoke Only in Questions

During an epileptic seizure, nine-year-old Roya lost consciousness for several minutes. The neurological cause remained unclear. There were no abnormal findings and no lasting damage. Nonetheless, something had changed. When she returned, Roya was not disoriented or confused, just quiet and still. When her parents spoke to her, she did not answer as before. She had no explanations, no stories—only questions.

"Am I back now?" she asked softly. "Was I someone—or was I something?"

Then: "Is light only there if you see it?"

And later: "Do we have edges, or only thoughts?"

Finally: "Am I the one who thinks or the one who listens?"

Over the next few days, the pattern remained. Roya did not ask about what had happened, but about what *is*. It was not childish curiosity, fear, or a grasp for control. She asked clear, unfiltered questions as if from a place that does not need answers.

Her parents were uncertain at first, although Roya did not seem unwell. Her behavior was not disturbed, only focused. Her speech was minimal but concentrated. Her eyes did not search, but they seemed to see, as if she had recognized something and decided not to name it.

In a conversation with a psychotherapist, she finally said:

"I think I was in a place where there are no sentences. Only truths. And they're like questions that don't get spoken."

When asked if she had felt afraid, she shook her head. "I wasn't alone. But I was no one. I was just a question mark. And that was light."

She began to observe things without wanting to name them. She used the word "I" less frequently. When someone spoke to her, she listened so attentively that even adults paused, not because they felt judged, but because they felt seen. Roya did not judge, but she did not miss anything either.

After a few weeks, everyday life returned. Roya's normal language returned. She laughed, she played, and she asked questions like other children do. However, something in her voice remained—a quiet spaciousness. Sometimes, when someone asked her what she had experienced, she simply said:

"Some things shouldn't be explained. Because they get smaller when you do."

Scientific Commentary

In Roya's experience, a form of consciousness emerges that is structured by questions. These questions are neither childish curiosity nor expressions of fear. The questions seem to arise from a liminal space, not from the solid ground of identity, but from an inner borderland where meaning is not yet fixed and the self is no longer centered. When she asks, "Am I the one who thinks or the one who listens?" she does not seek information. She gives shape to an experience in which the self is not a subject but a passage.

From a developmental-psychological perspective, language normally stabilizes experience by naming and ordering. Roya reverses this process. Her speech dissolves structure instead of forming it. For her, language does not become assertion but openness. In this, she approaches the modes of expression found in non-dual traditions, in which identity is understood as flowing awareness.

Particularly revealing is her statement: "I was just a question mark. And that was light." It is an expression of profound clarity that points to a release into a state beyond the need to be someone in particular. In both mysticism and transpersonal psychology, such experiences are described as moments of non-knowledge that are not empty, but luminous, without self-reference and yet deeply felt.

In phenomenology, especially in the writings of philosophers Edmund Husserl and Emmanuel Levinas, questions are not understood as a deficiency but as openness: ways of encountering what cannot be captured in concepts. Roya's questions are therefore not fragments of confusion but markers of a consciousness that moves without anchoring itself.

The default mode network, as mentioned previously, combines and distributes memory, language, word meanings, and stored knowledge to generate a coherent internal representation, including self-directed speech during childhood. After epileptic seizures, there may be an increase in inward-directed attention, which is presumably related to a temporary reorganization of the default mode network. Since children's self-concept is still developing, they have less awareness of their own thought processes and processing strategies. When uncertain, they tend to rely on exploratory questions rather than analytical explanatory models.

Liora

The Answer Before the Question Came

Liora was twelve years old when she lost consciousness for several minutes during an epileptic seizure. Although it was not her first, it was the most intense. Emergency services responded quickly, stabilized her, and brought her to the hospital. When she regained consciousness, she lay there calmly and entirely present. She opened her eyes, looked at her mother, and said only one sentence:

"There were no more questions. Because I had already felt the answer."

She did not speak of light, tunnel, images, voices, or beings. Yet what she described was just as powerful: a space without walls, no up or down and no direction. And yet it was unmistakably real. It was not a place one could describe, but it existed.

"I was no longer me," she said later. "I was what asks—before the question arises."

She did not speak of what she had seen. Instead, she spoke of what she no longer needed to know. It was as if everything had been present, not in words or thoughts but in a quiet certainty that needed no dialogue.

"It was like... when someone takes a deep breath, and you already feel what they're going to say, before they speak."

From the outside, little had changed. Liora went back to school, laughed, talked, and lived. Nonetheless, something inside her had shifted. Her parents noticed it first in small gestures, such as the calm way she answered and listened now, as if something had become familiar to her, something deeply close and long forgotten.

In therapy, she barely spoke of the experience, but she began to write down single words that described inner touches of lucidity, words such as "soft field," "return breath," "question body," and "answer light." She drew circles with dots, never exact and always pulsing, as if trying to show something that was meant not to be drawn, but remembered. When someone asked her what it was, she said:

"It's the inside of something that knows."

She never explicitly called her seizure a near-death experience. Instead, she spoke of a point within herself that "breathes when everything else stops." She did not call what she experienced soul, light, or spirit. Only:

"What remains when nothing asks anymore."

When someone asked if she was afraid of death, Liora looked thoughtfully out the window, then gently shook her head.

"I think fear always asks questions," she said. "But what we truly are already answers before we even hear them."

Scientific Commentary

"There were no more questions. Because I had already felt the answer." Liora's sentence does not follow the expected language of classical near-death experiences. There is no tunnel, no light, and no figure. And yet, her account points to an essential dimension of consciousness: The moment when the need for questions dissolves, through the presence of a deeper knowing. Her experience indicates a mode of consciousness that is not triggered by sensory input but that emerges from an inner presence. Liora does not describe an insight, a vision, or an explanation. She describes a certainty that is simply given. In her words: "I was what asks—before the question arises."

In some traditions of classical epistemology, knowledge arises in response to a lack. However, Liora's experience reverses this logic: The question never needed to be asked, because the answer was already alive. In phenomenological philosophy, especially in the work of Michel Henry, this is called *immanent evidence*, a truth that is not sought through intentional acts but immediately experienced.

Depth psychology also provides points of convergence. Here, archetypal fields are not to be understood as contents but as structures or forms through which meaning becomes visible. Liora seems to have entered such a field. She experienced what could be called *answer-consciousness*, a state in which the self no longer requires constructions but attunes to a truth that already sustains it.

In this context, her experience could be described as *implicit consciousness:* those forms of consciousness that are not at the center of a person's attention but still shape their experience, thinking, and actions. They are quiet, often preconscious processes that operate in the background. Implicit consciousness arises before language emerges in the form of images, bodily sensations, atmospheres, or resonances.

Quantum-inspired models of consciousness describe similar phenomena. Researchers such as physicists Menas Kafatos, Federico Faggin, and Amit Goswami interpret consciousness not as a product of neural processes but as a field of coherence in which information arises through resonance rather than transmission. In such a model, Liora's experience appears as a moment of synchronization to a knowledge that does not arise but is already present. In other words, she did not receive the answer. Instead, it already belonged to her.

REFLECTIONS

There are cases where it seems as if a child who survives a near-death experience speaks in a language without origin. Liva spoke words that belonged to no known language, and yet they sounded like home. Like the memory of a sound older than her first word. Her voice carried rhythm, structure, and depth, not from knowledge but from closeness. From a resonance that needed no language.

Perhaps language is not always what we learn. Perhaps it is sometimes what we remember when the self steps back and something deeper begins to speak. She found peace in the words. She had entered a space in which she made new and remembered words tangible at the same time. A language beyond nations, grammar, or cultural codes, and yet full of meaning. Perhaps that was precisely it: not information but a shared field of consciousness. It is possible that children who are not yet fully bound by the fixed structures of what is learned might have a natural access to this field.

Thus, Mark began to write poems to remember. His relationship to language was rooted not in communication but in being itself. He said, "Thinking has a different language. And I heard it once." Here, language might be understood not as a container but as a trace, as a gesture that points beyond itself toward silence. Mark spoke Old French as well as brief phrases in Greek, Sanskrit, and liturgical Hebrew.

For him, it was nothing extraordinary. No message, no miracle. It was simply natural. And therein may lie the crucial insight: language, in its essence, is not a tool but a relationship. A way of remembering an order that precedes the self.

What is a language that we do not learn, and yet understand? Can we be part of a thought without thinking it? Can there be communication without words? And what if language does not arise because we think, but because we remember? Perhaps Mark did not speak only syllables of a forgotten language. For a moment, he might have become a field of resonance for something that is always speaking but is audible only when the self falls silent. It is possible that he entered a language not made of letters but of closeness, of origin. And we might also belong to it, if we become still enough.

Some moments in children's experiences elude any kind of categorization. They cannot be reduced to grammar or narrative patterns, yet they are language in its most original form. Not born of thought but carried by memory. Not the product of the mind but the expression of a deeper, untouched source. Perhaps language, in its origin, is precisely this: an echo of the deeper layers of being rather than a function of the mind.

The "I" we take for granted often only emerges through language. We say, "I am"—and in doing so, we draw a boundary between inside and outside, subject and world. Yet when this "I" falls silent and is made permeable, another movement begins: language becomes space. No longer a tool of thought but an echo of being.

Children who have experienced a near-death moment seem, at least for a short time, to stand exactly in that space. They do not describe. They testify. Their words do not explain. They remember. It is as if their information comes not from the mind but from a depth where language no longer separates us. The deepest form of knowledge begins here: the realization that not everything needs to be said, because what is essential cannot be put into words.

Some children return from near-death experiences with silence. Roya returned with the ability to ask without insistence. Her words did not act as a sign of confusion but as a new form of clarity. To say, "Am I the one who listens?" does not mean doubt. It means remembering that the self is not the center but a lens. Her questions did not seek resolution. They resonated. Like open doors or light that does not need to shine in order to be seen.

What does it mean when a child does not ask, "What happened to me?" but instead, "Am I the one who thinks or the one who listens?" How do you speak to someone who does not need a self to be present? What remains when questions are open, and a world can be touched without being named?

Roya's questions are keys that strive not to unlock answers but to make presence accessible. Thus, her words do not appear as a deviation from development; they are an expression of extraordinary sensitivity. We do not need to name everything to recognize it. Sometimes it is enough to pause and feel the space between knowing and desiring. When was the last time you asked a question, not to learn, but to listen? Perhaps that question is already an answer.

Liora came back from a space that answered before the question was born. Her words were simple and yet unshakeable: "There were no more questions. Because I had already felt the answer." Perhaps such experiences are not about new knowledge but about a memory that is older than language.

Liora suggests that answers do not always come to us from a distance. Sometimes they are already within us, long before we dare to ask the question. And so what she brought with her from her experience might have been an attitude, a quiet certainty that carries weight without seeking to prove anything.

Perhaps some children return from near death so they can remind us that what we are, deep down inside, does not seek but already responds. Liora rarely spoke of what happened that day. Her way of living carried a clarity that needed no explanation. She listened more than she spoke. And when she did respond, it was often with a pause, as if the answer had already arrived, long before the question was asked. Liora did not analyze; she did not define. Her language remained gentle, careful, and clear. There was a certainty that was too complete to be explained. Perhaps this was her silent message: that the models of consciousness, which are so fixated on search and answer, miss the core truth.

The Silent Threshold

Near-death research often focuses on vision, with an emphasis on the commonly reported experiences of seeing light, tunnels, and beings. However, some children who return from a near-death experience do not speak of what they saw but of what they heard. These experiences belong to a rare but revealing category of threshold states, those in which perception does not follow the senses but arises from within. Survivors of brushes with death often describe coherence, emotional presence, and intuitive resonance, all arising without external triggers. And they describe a field of presence where sound is not needed for meaning to arise.

What is listening? Is it a function of the senses or a neurological event? Or is it a form of relationship between ourselves and a dimension that does not speak but answers? Perhaps what we usually call "listening" is only the outer surface of something far deeper. In contrast to everyday verbal language, which is designed to convey or explain terms or content, the language of consciousness may be resonant, constructed from connection, and designed to recognize.

The children in these stories often do not return with tangible details or concrete answers. They bring back a presence, a tone, a resonance. Their experiences remind us that not all truths require explanation or imagery. Some truths simply *are*, and these moments of recognition go beyond the intellect. They touch the core of our being with a silent, wordless reassurance that being near death can trigger an expanded state of being.

Matthias

Stillness at the Edge of Breath

Twelve-year-old Matthias was admitted to a children's hospice after a rare neuromuscular illness had slowed his movements and his speech. Each breath had become a conscious act. He rarely spoke, but his eyes saw everything. His gaze did not search, it received. He listened without reacting and with clarity.

In the third week, shortly after three a.m., the monitor beside his bed sounded. A nurse rushed in, but when she arrived, she found him without panic or pain—only silence. Matthias lay still with his eyes open, not breathing. His skin was warm. His body unstrained. As if he had stepped out of breath itself but stayed.

They began resuscitation. His pulse and his breathing returned. When they asked what had happened, Matthias did not speak of fear or confusion. He only said:

"I wasn't gone. I was in the stillness. And the stillness was gold." He described this as if it was not a message but a memory.

Later, he added: "I forgot to breathe, but I wasn't forgotten."

And: "It was like breathing without lungs. With my soul instead."

He did not describe any visions, tunnels, or voices. Only an expansion and a loosening. A golden stillness that held him. A light that was not bright. A silence that was presence and being.

In the days that followed, Matthias began to paint. He began with simple shapes: golden dots scattered across black paper, swirling spirals. One image he returned to again and again was a narrow, white rectangle bordered by darkness. He called it "the window." And when asked what it meant, he replied:

"It's the window the stillness looks through."

He did not explain. However, something about his words silenced the room. There was no interpretation, only recognition.

Matthias grew weaker, but he never seemed to shrink. He was not tired. He was clear. He was not between life and death, but deeper still: between breath and light, in a space not of endings but of seeing.

Three weeks later, Matthias died quietly and without any signs. The nurse found him early in the morning. His hand lay open, palm upward, and he wore a faint smile on his face, as if he had recognized something. The stillness. The gold. The window through which the stillness had seen.

Scientific Commentary

Central respiratory depression often occurs shortly before critical states in which consciousness becomes severely altered or begins to fade. As the moment of death approaches, the respiratory drive within the brain is diminished or completely absent. It is not the breath that ends but the need for it. From a neurophysiological viewpoint, such states of deep transition are difficult to explain. When breathing stops and oxygen drops, neuronal activity declines rapidly. And yet, people report moments of great clarity and awareness. This suggests a form of consciousness not entirely tied to neural function.

In mystical traditions, breath is more than biological. It is the threshold between body and awareness. Rhythmic. Voluntary. Symbolic. Breath often becomes consciously perceived, suspended, or dissolved in moments of transition. In deep meditation, the breath recedes from the foreground of awareness into the background, sometimes almost imperceptible.

In palliative research, people on the verge of passing often describe the moment of the last breath not as frightening but as a return to something familiar. Tibetan teachings describe the final breath as a drawing inward. The image of the window—stark and minimal—is deeply symbolic. In depth psychology, the window often appears as a boundary between the seen and the unseen, between presence and absence. However, Matthias reverses the frame. The stillness is not what he looks through. It is what looks back.

Noémi

The Silence That Sounded

Noémi was ten years old when she collapsed in her classroom due to a severe allergic reaction. Her breathing stopped and her pulse nearly disappeared. Her body folded in on itself.

She was successfully resuscitated. When she awoke, she was alert and responsive, but something in her presence had changed. Her eyes looked as if they were listening. She did not seem confused or withdrawn. She was deeply present, as if she had heard something too vast to describe. At first, she said nothing. There was no fear or uncertainty in her. Only a quiet, open presence, as if her attention had shifted.

Days later, on a walk with her grandmother, she said:

"It wasn't silent there. It was the silence that sounds."

She did not try to explain it. Words seemed too narrow. When she did speak, her sentences rose from within, in the sense of memory.

"I didn't see any light. But what was there had no darkness."

And: "I wasn't in a space. I was what gives space."

Then: "There was a tone. But without sound. And the tone was me."

Her parents were both moved and unsettled. No tunnel. No light. No otherworldly encounter. Instead, the girl spoke of hearing a soundless tone. And yet Noémi did not seem detached, just quieter and more grounded. She was deeper in the now. Her teacher wrote that she spoke less, but her responses came "as if from a different angle." A friend said, "Sometimes she just looks into the wind. And you know she's listening."

Noémi began writing down her thoughts, single lines that were scattered on sheets, tucked in coat pockets, or slid under pillows.

"I don't listen with my ears. I listen with the part that knows when someone is sad." Or: "Some people talk too much because they can't hear what matters."

She never spoke of God or of a place, not even of consciousness. She only spoke of one thing:

"It is there. And it has no language. But it calls. Very softly."

To this day, she says little about what she experienced. But when asked, she sometimes answers with a quiet gaze or a single sentence:

"I think silence isn't a lack. It's a beginning."

Scientific Commentary

Most research on near-death experiences focuses on visual phenomena: tunnels, familiar people, or life reviews. However some of the most subtle reports come not through sight, but through presence. Time and again, children describe a kind of hearing that has no source, no tone, or no direction, and yet it is unmistakable. The children do not describe acoustic events. They speak of an inner contact. A form of listening directed toward something that does not speak in words but is completely present. Noémi's words capture this perfectly: "I don't listen with my ears. I listen with the part that knows when someone is sad." What she describes is resonant awareness.

In transpersonal psychology, such states are understood as expansions of consciousness. Listening, in this context, is not an act of perception but of participation. Consciousness attunes itself to the field of experience like an instrument that begins to resonate.

Neuroscience only cautiously approaches such experiences. Studies on deep meditation, *samadhi* states, and near-death awareness show that when external sensory input fades, inner networks tied to emotion and meaning remain active. People report hearing tones, sensing voices, and feeling coherence, without any external trigger.

These experiences are not chaotic. They follow an internal logic shaped by memory, emotion, and something else: a kind of inner attunement. The above-mentioned *samadhi* is a concept from yogic, meditative, and contemplative traditions of India and refers to states in which attention becomes so unified and inwardly absorbed that the usual boundaries of self, thought, and perception temporarily dissolve.

Dalia

The One Who Waved

Dalia's illness brought her to the edge of life when she was only five years old. A congenital heart defect, after several surgeries, had left her small body in a fragile rhythm between hope and exhaustion. Her limbs grew weaker, but her gaze became more awake than ever. She spoke rarely. When she did, the room listened. That happened not because her voice was loud, but because it carried something that needed no volume.

On the third night after a sudden circulatory collapse, Dalia lay quietly in her bed. The monitors showed her body was failing. As a nurse held her small hand and hummed a lullaby from long ago, Dalia slowly opened her eyes. She looked toward the ceiling and smiled. Then, gently and with unmistakable clarity, she lifted her hand, not aimlessly or weakly, as if greeting someone. She did not wave to someone in the room but to someone beyond it.

"There's someone there," she whispered. "She is beautiful. And she's waiting."

Dalia never said who she was. There was a certainty in her voice that silenced all doubt. In the following nights, the moment returned, always just before sleep. Her vision was always accompanied by the same movement: a slow, deliberate raising of the hand, as if she could see what others could not. And afterwards, that quiet smile which was awake and present. It was as if one part of her already stood in the light while the other part waited patiently beside the bed.

Sometimes she murmured things like: "I'm not finished yet. But I'm allowed to look." Or: "It's like a light that breathes. And it knows my name."

She never spoke of her visions. She did not describe places or figures. She did not ask questions. She simply waited with the gentleness of someone who already understood.

The day before her death, her mother stayed by her side. Dalia was tired but calm. She spoke little. Just before drifting off, she laid her small hand on the blanket, looked at her mother with great clarity, and said,

"If I wave, you don't have to be sad. That's my yes."

That final night was quiet. There was no struggle, no fear. In the half-lit silence, Dalia's hand lifted one final time in a gentle and natural way, as if passing through a door that no one else could see. Then her hand came to rest softly on the pillow as she took one last breath. She was gone without any haste or resistance.

Scientific Commentary

Dalia's case belongs to what are known as *deathbed visions*, experiences that occur shortly before death in the form of perceptions, dreams, or visions. Unlike near-death experiences, in which individuals come close to death and subsequently return to report their stories, deathbed visions describe experiences that arise during the very process of dying, usually without a return to life. But they are like near-death experiences in that they offer a cautious glimpse of what might await us after death.

People often report seeing deceased relatives, friends, and pets as well as religious figures, luminous phenomena, landscapes, or otherworldly spaces. In addition to visual perceptions, some people also report auditory, olfactory, tactile, or intuitive experiences. Auditory experiences may include calming voices, melodies, or clear messages. Tactile and intuitive experiences often involve a sense of "energy" or "presence" in the room or the feeling of "being accompanied." Such occurrences are often marked by a sudden sense of peace and comfort for the person who is in the process of dying. For this reason, some researchers prefer the term *death-related sensory experiences.*

In the field of palliative care, the dying process is often described as peaceful, with the dying person passing away with a calm smile. Relatives or caregivers sometimes report that patients suddenly look toward a specific point in the room and appear unusually attentive shortly before death.

These patients might be experiencing the phenomenon of *terminal lucidity,* where severely ill or cognitively impaired patients become suddenly clear, alert, oriented, and communicative shortly before death—sometimes for minutes, hours, or more rarely, one or two days. This unexpected clarity is often in sharp contrast to the person's condition during the preceding weeks or months.

REFLECTIONS

There is a form of listening that requires no ears. It is a perception that comes from within. And yet, some of those who returned from near-death describe not sound, not a voice, not a tone in the ordinary sense. It is a tone without sound. A silence that resonates, which arises at the edge of perception where self, identity, and vibration no longer follow the rules we usually trust.

What if listening begins where sound ends? What if silence is not an absence to be filled but a presence to be received? The silence Matthias and Noémi carried was a profound fullness, a reminder of the deep interconnectedness of all life. What remains from such experiences is not a new set of beliefs or teachings, but a shift in perception. Sometimes, the most powerful transformations are the quietest.

Matthias' experience touches a form of near-death encounter that is often overlooked. What Matthias describes is pure presence: a state of complete breathlessness, with a sense of being held in stillness. In Matthias' case, the striking part is that breath did not "fail" but was, as he said, "forgotten." This phrase suggests not a loss of control but rather a gentle release. When he speaks of a "golden stillness," he does not mean nothingness. He means a space. This description appears in other near-death accounts as well, especially from people who report conscious awareness during full cardiac arrest. They describe neither darkness nor chaos but a quiet presence. A stillness that holds.

What Matthias experienced was a state of being in which breath ends and consciousness continues to breathe. His words: "I forgot to breathe but I wasn't forgotten," might express a dissolution of the body that leads to relation. Something remains. And that something is awake. Silent. Perhaps in such quiet and profound experiences lies a clue that consciousness does not vanish but remembers. Not with words, but with the gentle certainty of a child who looks back and says: "The stillness held me."

How is it possible that a child without oxygen, without measurable brain activity, sees nothing, hears nothing, and yet remembers? How do you describe a stillness that holds? Why does a twelve-year-old not say, "I stopped breathing," but: "I forgot to breathe"? And who or what is it that holds us when breathing ends?

Matthias did not experience a loss of breath; he experienced its release. Who are we when even breath falls away? What remains when there is no function to sustain us, but presence continues? Perhaps in such moments it is not about what happens, but what remains. There are moments when life does not end but softens. Not abruptly. Not dramatically. Like a breath that is still there but no longer asks for anything. Perhaps the true threshold is not where the heart stops but where the self dissolves into something that no longer asks. When a child like Matthias returns with drawings instead of explanations, with windows instead of visions, he brings no message. He brings a state. A sense. A form of knowing.

Noémi did not return with a story. She returned with a tone, not one to be heard, but one to be remembered. And perhaps that is the most faithful kind of testimony. A resonance. A stillness. A beginning that never needed sound.

Noémi did not describe a tunnel, a light, or a voice. What she encountered was a presence that did not speak but held. Her story is not about what was seen, but about what was sensed when everything else became still. Noémi's account represents a form of perception rarely studied: soundless listening. This is a mode of knowing that transcends ordinary auditory function. Noémi describes a tone without a source, a kind of listening that is not sensory but relational. "The tone was me." In mystical literature, silence is not understood as a lack but as depth and presence. Noémi's experience reflects this; she does not perceive silence as separate—she becomes it.

"It wasn't silent," she said. "It was the silence that sounds." This is not metaphor. It is the description of a conscious space beyond acoustics where listening no longer depends on sound but on significance. So perhaps that is the essence of her story: that listening does not begin in the ear but in the part of us that responds when nothing is said.

Dalia's story is one of the most subtle yet powerful cases. She offered only a gesture. And yet in that one quiet wave, a form of knowing was made visible that reached deeper than any sentence. It was not a coincidence or reflex. It was intention. Meaning. A knowledge that revealed itself without any words. Such gestures belong to the most subtle forms of experiences in the immediate proximity of death in childhood.

Unlike adult deathbed visions, where dying people often report having contact with deceased relatives, end-of-life visions in children tend to involve an unknown presence. The nurse called it a "greeting." Dalia's mother understood it as a passage.

Dalia's words reinforced this silent action: "There's someone there. She is beautiful. And she's waiting." No name. Just a presence that needed no justification. When she said: "It's like a light that breathes. And it knows my name," that was not a metaphor. It was a recognition.

The kind of relational, immediate, and unambiguous knowledge that went beyond images.

Her final gesture, "If I wave, you don't have to be sad. That's my yes," carried a clarity that went beyond farewell. She described not letting go but being received. There was peace in this gesture, for her and for those who remained.

Perhaps what Dalia saw is less important. Perhaps it is enough that she saw. That she smiled. That she waved. Her wave explained nothing. It affirmed. Quietly. Completely. And perhaps this is precisely what all words seek to express: a presence that sustains and a "yes" that needs no sound.

Perhaps death is the great becoming-still, in which the bond does not end but expands. A window through which we do not look out but through which something looks back. Quiet. Recognizing. Perhaps that is the greatest grace: that in the end, there is no word but a breath no one hears except stillness itself.

When the World Feels Different

In the stories we have heard so far, children have crossed thresholds, and each story has brought us closer to understanding the transformative power of these experiences. Yet it is not always the view of life and death alone that changes. Sometimes the very fabric of sensory experience itself undergoes a metamorphosis that touches both mind and body.

In this chapter, we turn to those children who, after their return, perceive the world in a new way. These are not "new abilities" in the narrow sense but rather another way of sensing reality. Their senses seem heightened, more permeable, differently attuned. Through this, the world itself appears transformed.

The return from a space of stillness or transcendence into a world full of hardness, harsh brightness, and urgency is palpable and unsettling. In such moments, it is not the world that has changed, but the person who returns. Sound feels sharper, light more artificial, language hollower, and touch more abrupt. This might be the result of a recalibration.

Mira

The Return to the Wrong Light

Mira was eight years old when her body gave way to a severe allergic reaction. Her lungs closed, her blood pressure collapsed, and her heart slowed. For minutes, she had outwardly disappeared. By the time the ambulance arrived, her heart had stopped. The doctors worked desperately to revive her and it took nearly ten minutes of intense resuscitation before she was brought back.

Mira later recalled that when she opened her eyes after those ten minutes, the light was *wrong*. It was neither terrifying nor dark, just wrong. It was too white, too sharp, and too loud. It was not like the light she had just been in, which was soft, warm, and the kind that surrounds you without touching. The "new" light was different. It stung, pulsed, and it buzzed through the walls, through the ceiling, and through her skin.

When she returned, it was not relief she felt. It was overexposure, a kind of sensory overwhelm that had no origin.

"Everything was too loud," she said later. "The light, the voices, even the air had noise."

The days that followed were difficult. She hid under blankets and whispered when she spoke. She avoided sunlight and closed her eyes when the refrigerator whirred. Even silence, it seemed, had gained weight. Her parents thought she was overstimulated. The doctors called it *post-hypoxic sensory hypersensitivity*, which is a reaction of the nervous system after critical oxygen deprivation. However, Mira did not think she was sick.

"It's not like I have more senses," she said. "It's like the world doesn't have filters anymore."

Fragments of what happened during her collapse came slowly. She spoke of a space filled with golden quietness, a movement that felt like floating without direction, and an embrace that did not come from arms but from everywhere. She said there was a color that had no name and a feeling of being held—not touched, just known.

"It was light," she said, "but it wasn't bright. It was quiet but not empty. It didn't come from outside. It was everywhere."

Her return felt like falling into intensity. Everything in this world was harder-edged, narrower, and faster. Mira did not say she was frightened, only misplaced.

"It's not the world that's different," she said. "It's just that now I hear how loud it is."

Over time, she found ways to cope. She wore soft fabrics, moved slowly, and learned which corners of the house had the least light. Her voice grew quieter and her eyes learned to search for shadows. She began to describe sounds that had color, fabrics that had tone, and silences that had weight. Her parents were worried. The doctors suspected *synesthesia*, a special form of perception in which a stimulus in one sense is automatically accompanied by a second sensory impression. Mira disagreed.

"It's not mixed," she said. "Things go through me differently now."

She never asked to go back, but she also never called it a dream. What she experienced did not have a name or a shape, just a sensation.

What stayed with her was not the golden light, but the contrast it created. It was the first look at the wrong light she remembered because everything else, for a moment, had been right.

Scientific Commentary

Mira's story opens a lesser known but increasingly acknowledged dimension of near-death research: the transformation of sensory perception after a profound threshold experience. While many accounts focus on light, peace, or encounters with the beyond, only a few explore what the return feels like, more specifically, how the world sounds, smells, hurts, or shines when a person comes back. Sometimes, the world feels too loud, too sharp, and too much. Light becomes harsh, sounds pierce, and texture overwhelms. It seems that the nervous system returns from another mode of reality, without its familiar filters.

Clinically, these states are often labeled as *sensory hypersensitivity*, a reaction of the nervous system that can occur after critical oxygen deprivation. Senses react more strongly than before the hypoxia, but they are not linked to each other. People usually experience it as stressful.

What stands out is not only the intensity of her perceptions, but also the integration of this sensitivity into her way of being. Mira expressed sounds as colors, silence as heaviness, and fabrics as tones. This state of expanded consciousness, called *synesthesia*, can accompany sensory hypersensitivity. Synesthesia can also occur after a period of reduced oxygen supply to the brain which leads to a desynchronization of sensory networks and disturbances in stimulus processing. Certain brain regions become hyper-excitable, particularly in the visual and auditory cortices. Furthermore, the sensory filtering functions of the thalamus are altered so that sensory impressions overlap.

In contrast to people who have sensory hypersensitivity, however, many of the people affected by synesthesia, particularly children, do not describe their condition as a deficit. On the contrary, they feel opened. Children, still unconstrained by adult expectations of "normal" perception, often describe this state without fear, complaint, or distress.

Spiritual traditions, such as Zen or the yogic practices of samadhi and Kundalini, also know of this phenomenon: the reentry shock. According to Stanislav Grof, a psychiatrist who specialized in transpersonal psychology, the "shock" does not arise from the extraordinary experience itself but from the contrast between the expansive inner reality and the return to the densely structured state of everyday consciousness. This state should be understood as a normal adaptive response of the nervous system, which might require a gentle reintroduction to external sensory stimuli.

Elin

The Child No One Could Touch

Elin's heart stopped beating when she was nine years old. A severe allergic reaction had swept through her body with terrifying speed and caused her circulation to collapse and her lungs to fail. For several minutes, there was silence. Resuscitation was successful; medically, she was stable. However, something had changed.

Elin returned in a conscious, alert but distant way, into a silence that seemed to come from somewhere far deeper.

"Words were too loud," she later whispered. "They broke the delicate things."

At first, she refused to speak—not because she could not, but because language itself felt too sharp and too dense. She flinched from voices and recoiled from touch, allowing no handshakes or embraces, not even a comforting gesture.

"It scraped," she said. "Not on my skin. Deeper."

Instead of speaking, she drew pictures with astonishing calm and precision. She drew overlapping colors that felt like moods, rooms without angles or perspective, forms without direction, and lines that drifted like breath across the page. Her drawings did not depict scenes. They conveyed states: feelings without names or spaces without walls.

Her parents were confused, her teachers unsettled. Only one caregiver stayed close without pressing; instead, she placed soft fabrics, translucent paper, and fine brushes beside her chair and waited. Elin responded, and a quiet form of trust began to grow in the stillness. Only much later, and only in fragments, memories of what she had experienced began to surface.

"There was a place," she said. "Everything was soft. But not like pillows. More like... nothing had edges."

No sounds. No walls. No voices. And yet everything was there.

"I didn't have to speak. I just was."

And then, when she suddenly returned, "It was like falling into noise," she said. "Like crashing into language."

The light felt harsh. The air was too full. Words, once invisible bridges between people, now sliced through her like blades. She was not afraid, but she was porous.

"I was too thin for this world," she once said. "Not my body, my inside."

It took months before she began to speak again. And when she did, her words came gently, as if they were feeling their way forward. Listening had become her new language. She rarely asked questions and her sentences sounded like echoes from behind the veil.

"I think I came back into something that no longer fits me," she once said.

And still, she stayed quiet, attuned, and present. When the world became too loud again, she would simply say:

"I need to paint now. It's gotten too sharp again."

Scientific Commentary

Elin's experience opens a field of perception rarely articulated. She returned with a changed level of sensitivity, an altered mode of sensing that language cannot convey. What she brought back from the threshold of death is a state. A new kind of awareness, and a silence too fine for the world to hold.

Children who remain silent after a near-death experience often describe words as too loud, too angular, and too small. Language distorts and fractures their experience, misrepresenting what is too complete to name. The withdrawal is then not an expression of fear but of fidelity to something more subtle.

From a clinical standpoint, her response might be categorized as *post-traumatic hypersensitivity*, a temporary state of heightened sensitivity after extreme stress, which can lead to sensory overstimulation, dissociation, or trauma-induced mutism.

Neuroscience indicates that threshold states, such as hypoxia or loss of consciousness, can trigger changes in cortical plasticity. Filters dissolve and sensory impressions enter unbuffered. Some researchers compare these states to forms of "raw perception" seen on the autistic spectrum: an openness to stimuli without protection.

However, for children like Elin, things are different. They do not lose connection to the world. They experience it more subtly, more openly, without the usual defenses. They sense what others cannot. And that is what makes the return so difficult, because they can no longer exclude the world without losing something essential.

Elin's drawings are traces of direct visual knowing. Her art is not avoidance, it is communication beyond language. In consciousness research, there is growing recognition that some near-death experiences cannot be contained within narrative forms, but only conveyed through atmospheres, states, and interactions.

Catherine

The Light Beneath the Water

Six-year-old Catherine nearly drowned in a public swimming pool. No one had seen her drift into the deeper water and by the time she was found and pulled to the surface, she was unresponsive. Her small body had begun to shut down. Oxygen deprivation was severe, and she lost consciousness.

Catherine later described a memory of light, stillness, and peace. She recalled seeing herself from above: her body floated motionless while adults moved frantically around her. The scene felt strangely distant. Her own body appeared unfamiliar to her, almost like an empty garment. She did not feel fear or confusion. She was aware of the scene without being bound to it.

"A slender thread of light," she said, still connected her to her body, but she no longer felt confined within it.

Then, from the periphery of perception, beings began to appear, namely forms of light, translucent and softly layered. They were not human in any ordinary sense, and yet not unfamiliar, either. Colors shimmered within them like flowing currents: rose, gold, pale blue. Though they did not speak, Catherine understood them completely.

"The love they gave didn't need any words," she later said. "It was complete. Unconditional."

One of the beings turned directly toward her. It did not use language, but its presence carried a message that required none: *Everything is good. You are loved. You are safe.* There was no judgment, instruction, or demand—only presence and peace. Catherine experienced this moment as a return to something deeply familiar, though long forgotten. With that recognition, something shifted inside her.

When she returned, the moment of transition was not abrupt but soft, like slipping into warm light and finding oneself breathing again. From then on, she spoke with calm certainty about life continuing. To her, death was no longer an end but a homecoming. Besides having a different perspective on death, she also had a different view on life. Those around her noticed the change.

Catherine, already a gentle child, now carried an unusual depth. She seemed more attuned to things that others overlooked. She noticed pain, especially in animals, small children, and the elderly, and with a clarity that went beyond empathy. It was as though she could feel what others could not say. She was grateful for little things, such as the golden hue of morning light, the scent of lavender on a blanket, the hush before sleep.

Her parents began to call her "quietly wise." There was calmness in her and patience not typical for her age. She spoke more slowly and paused before answering. And sometimes, she would simply listen as if tuning in to something just beyond the moment. It was as though she carried trust larger than herself.

Scientific Commentary

Catherine's experience occurred during a state of *acute hypoxia*, a dramatic deprivation of oxygen to the brain that typically leads to unconsciousness, disorientation, or irreversible damage. From a neuroscientific perspective, no coherent perception would be expected, let alone a structured memory. While sensations such as visions of light, feelings of floating, or time distortions can be induced under laboratory conditions through electrical stimulation of specific brain regions or by psychoactive substances, these reactions are generally considered fragmented, distorted, or hallucinatory.

The memories she brought back were not chaotic, fragmented, or transient. She didn't experience euphoria. Instead, she remembered clarity, structure, and coherence. She felt inner peace. This quality clearly differs from the typical effects of cortical disintegration or hypoxic hallucinations as frequently documented in emergency medicine.

What is particularly remarkable is the nature of her perception. She does not describe scenes shaped by religious belief or cultural conditioning. There is no heaven, no saints, and no narrative structures of an "afterlife." Instead, she recounts simple, immediate impressions, such as light, silence, a view of her own body from a distance, and beings of light who did not need any words. This kind of experience is not unusual. In many documented cases, exactly these motifs appear: out-of-body awareness, nonverbal communication, and a deep sense of peace.

Magda

Three Journeys into the Light

Magda suddenly fell seriously ill when she was eight years old. What began with mild fever and fatigue escalated rapidly. Within days, her condition deteriorated. The diagnosis was a rare systemic infection with an unpredictable course. By the time she was transferred to the intensive care unit, her body was already beginning to withdraw. She was placed in an induced coma, intubated, and continuously monitored.

For nine days, she showed no outward signs of consciousness. Nevertheless, she later said:

"Something was happening within."

When she spoke again, her calm and clear words came without hesitation from a depth no one had expected. She spoke of three journeys, three distinct experiences. Each had its own atmosphere, imagery, and emotional quality, yet all were connected, like gates opening onto the same vast design.

In the first journey, she saw her deceased grandmother. The occurrence was suffused with light that was familiar and still. Her grandmother appeared just as she had looked in old photographs, only brighter, more vivid, surrounded by a gentle glow. There were no words, no explanations. Magda described it as a silent reunion filled with love, as if a bond that had never truly ended had become visible again, just for a moment.

The second journey brought her into a realm of color and sound.

"Everything was singing," she said.

Colors moved like melodies; tones shimmered like light. There were no boundaries, no separations but only spaciousness suffused with harmony. Her senses felt opened, transparent, and connected.

"It was brighter and more real than being awake," she said. "Everything was there, but without weight."

In the third experience, she found herself in a meadow. Sunlight poured across wildflowers, and children were laughing and playing in the grass. Some turned toward her and smiled, as if they had always known her.

"The light was friendly," she said. The air felt "soft and round." There was no purpose, no pressure—only presence.

After waking up, Magda was no longer the same. She spoke softly and steadily of an inner peace that had moved into her chest and stayed there. Her fear of death was gone. When she spoke of farewell, it was not with heaviness but with a kind of lightness.

Those around her noticed changes, too. She listened differently. Responded more gently. Her words came more slowly, more deliberately, as if something within her had widened. Each of these experiences remained distinct in her memory. They did not blur or dissolve with time. They left traces in Magda's life that did not disappear.

Scientific Commentary

Near-death experiences can also occur under deep sedation, such as in an artificial coma or under general anesthesia, during life-threatening situations. However, research suggests that the likelihood of remembering such an experience is reduced under the influence of sedative medication. Children in comas are often regarded as neurologically inaccessible, not only in terms of immediate responsiveness, but also when it comes to retrieving conscious experience after the event.

Since the 1990s, an increasing number of documented case reports have described children's experiences that are coherent, emotionally charged, and strikingly consistent across cultures. A classic near-death experience can be understood as an inner awareness accompanied by a separation from the external world, representing a condition of disconnected consciousness. In comparison, states such as meditation or fainting are typically experienced at different degrees of wakefulness and connectedness.

From a medical standpoint, a nine-day coma is a severe and exceptional neurological state. During such a condition, brain activity is drastically reduced. Sedation, systemic inflammation, and reduced oxygen supply heavily suppress brain activity. Cognitive functions such as episodic memory, time perception, and self-awareness are compromised. An induced coma also might involve the disconnection of consciousness, as mentioned above. Conventional neuroscience assumes that under such conditions, at most, fragmented hallucinations arise that are transient traces of uncoordinated activity or chaotic reorganization processes during reawakening.

Magda's experience did not conform to that model. Her experience was neither fragmented nor chaotic. It appeared self-contained, structured, and carried by a quiet inner order. The elements she described, encounters with the deceased, spaces of light, color, sound, and a peaceful landscape, are consistent with many independently documented near-death experiences. This recurrent pattern across all age groups, cultures, and worldviews suggests less of an individual fantasy and more of a shared memory structure, something that appears archetypal and is experienced on a deeply personal level. What distinguishes near-death experiences in children from those reported by adults is usually the absence of a socially constructed, meaning-driven, or religious framing of the experience.

REFLECTIONS

Not every return brings relief. Some children come back from the edge of life physically stabilized and medically healed yet inwardly transformed. The experience beyond the threshold does not bring peace but intensity. They re-enter the same room, the same bed, and hear the same voices, and yet nothing feels familiar. The world is still the world, but it no longer fits.

These kinds of post-threshold sensitivities may point to a reconfiguration: A nervous system that has encountered stillness or vastness and now returns without its previous protective layers. It is not always easy to endure. It reveals a deeper attunement, an immediacy of presence that has less to do with fragility than with another form of being present. Perhaps some children return from stillness with a skin that "listens" more deeply. What they need then is not numbing but space, not correction but kindness.

Often it is not the experience itself that lingers, but the contrast. The harshness of a light that once seemed warm. The sharpness of a sound that had until then gone unnoticed. The density of everyday life after the expanse of something different, something softer, vaster, quieter. For Mira, it was not the fear of what had happened, it was the quality of what she returned to. She did not cry. She did not protest. She moved more slowly. She listened more deeply, as if every movement now touched something more subtle within her. She simply said: "Everything was too loud."

Some call it sensory overload, but it might be something else: a shift, a tuning. An openness that no longer filters and that does not push anything away. Every sound becomes a wave. Every object has a presence. Every shadow an invitation to listen inward. Mira was not weaker; she was more permeable. And this permeability carried its own quiet strength.

Mira carried back an increased sensitivity: a transformation in the way she perceives light, colors, and the world itself. This change is quiet but profound. It suggests that near-death experiences may alter not only a child's view of life and death, but also the very way in which their senses relate to the world. Her story points toward the possibility that such sensitivity is not unique, but something we can all glimpse in certain moments. What Mira experienced becomes an invitation: to remember what it feels like when perception itself begins to change.

This raises the question: What happens in the body during such an experience? Can death or the state of being near death truly shift sensory perception and open new pathways to the physical world?

For Elin, words had become too dense and touch too abrupt. The world felt sharply outlined. Her withdrawal did not arise from fear but from loyalty to something that had touched her, something more subtle. Her drawings became a way of being, a place where she could remain attuned. An alignment that does not seek to be understood, only received. What she brought back was a gentle and complete state. Her silence was not absence but presence in its purest form. She was transformed in a way that affected her entire being.

Elin's refusal to speak was not a withdrawal but integrity. She had been at a place where edges no longer mattered, where sound had texture, and where being was enough. And yet, Elin did not lose her words. She simply perceived them as too coarse. Her gestures became her voice, her drawings her testimony. What she experienced was a tuning. A refinement. A state beyond narrative. And this state did not fade. It continued, in her silence, in her rhythm, in the way she noticed the air before it moved.

What she remembered was not what she saw, but what she was. A texture too subtle for her to describe and too true for her to forget. Elin's near-death experience brought forth a quiet yet profound transformation in her relationship to life. She did not seek comfort in language or familiarity. She moved within another kind of peace, one that lives in silence, in colors, in space.

Catherine carried a trust larger than herself, one that had been given to her as an experience. She described the pain of others with a clarity that goes beyond empathy. For her, compassion seemed to arise as a deeper mode of perception. She sensed what others could not articulate.

The peace that settled in Magda after her three journeys beyond the threshold between life and death felt like a newly learned fundamental tone. Her slower way of speaking reflected an expanded inner landscape. Words were no longer used as quick tools but as something that was placed with care. To adults, this often appears as unusual maturity, yet it is less a matter of growing older than of having entered a new relationship with the meaning of life.

What remains when the noise of life has faded is not emptiness but a wave of connection. A quiet hum that carries the memory of what we once knew, before we learned to speak, before we began to divide. It is the hum of a world that does not need us to understand it, only to feel it. And perhaps, if we allow ourselves to linger in silence, we may also hear it again—that deep, eternal pulse that binds us all.

In the Liminal Space

There are experiences that cannot truly be told, because they are not incidents. They have no plot, no setting, and no sequence. They happen in a space where the world itself becomes blurred, where perception no longer separates, and where inside and outside lose their direction.

This space eludes description because it has too little form to be named. It exists between thoughts, between breaths, between identities. It is a transition that knows no destination. A state that is remembered not as a scene, but as a quiet, fundamental shift. Children who touch such in-between spaces describe a quality of presence, a being with no center, a closeness without direction. They speak of something that is like a field in which experience itself becomes possible.

These experiences can be the most unsettling. They reveal nothing we can hold on to. And yet, it seems that precisely in this indeterminacy lies a particular form of precision. Not every experience presents itself as an image. Some experiences do not leave a memory in the usual sense, but rather a changed stance toward the self.

The following accounts approach the same in-between space from different directions. Each of them moves toward an indescribable state that can only be intimated, more felt than understood. They do not unfold through interpretation, but through a quiet accompaniment.

Milos

The Moment In Between

Milos died for the first time when he was eight years old. It happened during an open-heart surgery. Complications arose, and the monitors went silent. There was no pulse and no measurable brain activity. Five minutes of clinical stillness. And yet, what Milos later described was not stillness. It was a moment suspended—neither here nor gone.

"I wasn't away," he said, "but I wasn't here anymore, either."

He did not speak of tunnels, voices, or light. He spoke of motion inside silence and of something like thinking without direction. There was no space and no body. He described a presence which was denser, deeper, and more real than anything he had known before.

"It was like a single point," he said, "but everything was there, all at once."

Scientific Commentary

Milos' account challenges the prevailing assumptions not only of near-death experience research but of neuroscience and philosophy itself. He does not report visions or voices. He speaks of a kind of presence that is not sensory, not imagistic, and not narrative. In clinical terms, this poses a paradox. His brain showed no measurable activity. And yet, what he describes is not unconsciousness but intensified awareness. It is not confusion but rather coherence.

Coherence is central to the idea developed by consciousness researchers Stuart Hameroff and Roger Penrose, which proposes that consciousness interacts with the brain through quantum coherence. This interaction may occur specifically as high-frequency oscillations in the neurons' microtubules embedded in quantum-mechanical processes.

The separation from the physical body is often associated with a deep sense of peace and tranquility that is experienced as an inner silence, in which time and space are dissolved.

In quantum physics, the phenomenon of nonlocality is no longer a fringe idea. It is a repeatedly confirmed principle: two entangled particles can remain instantaneously connected across any distance. They respond as one system, regardless of time or space. No signal passes between them, and no exchange of classical information occurs. Nevertheless, they "know" each other. This challenges our notion of causality and the concept of consciousness as a local storage unit.

Adrian

The Breath Between Worlds

Adrian was almost fourteen when a severe case of pneumonia placed him in a life-threatening condition. The illness progressed rapidly, marked by a rising fever and increasingly shallow respiration. His sleep became restless. Then one night, at approximately four a.m., he lost consciousness. In the hospital, his respiration failed and his oxygen levels dropped abruptly, causing his body to stop responding for several minutes. What happened in between was not shown on the monitors.

When Adrian woke up, he was calm. He was awake but not fully "back." It took three days before he spoke. When he finally did, it was not a report. It felt more like an echo resonating through him.

"I wasn't dead," he said. "But I was somewhere else."

He described a space that was not a space in the ordinary sense. There were no walls, light, or images—only breath. The breath was not his own but that of the world. The air, he said, was vibrating, not as wind but as tone.

"I was between two breaths. And everything is possible there."

He did not say much. When he spoke, it was as if time slowed, and as if a new kind of space opened—a space of transitions, as he called it.

"There was something getting ready. Not me. But what comes."

He made clear that he was not alone, although he did not see anyone. It was only a feeling; indeed, several. There was a presence within them that he recognized, though he had never met the presence before.

"He knew what I would feel, before I felt it. And I knew that he knew."

It was neither memory, nor knowledge, nor a vision. It was a mutual familiarity without a past.

"I think I was in something that prepares time," Adrian said. "Not like a clock. But like a movement that is already breathing you."

After his return, he began to write. The pieces of writing were lines and soft fragments, each like a breath held in silence. One sentence returned again and again.

"I wasn't there. But it saw me."

Scientific Commentary

Adrian's experience belongs to those rare cases that stand out because of their inner quality. He does not report a light, a life review, or an encounter. Instead, he recounts a state beyond the usual patterns. A space that is not defined by images but by sensation. It is precisely this simplicity that makes his experience significant.

According to some neuroscience theories, consciousness is present in the entire body. Through inhalation and exhalation, consciousness is integrated. Oscillations in the form of shifts between resting states and electrical tension across cell membranes are coupled to inhalation and exhalation. Therefore, breathing plays a crucial role in the human subconscious and in perception.

Adrian did not speak of the past, but of what was to come. "There was something getting ready. Not me. But what comes," he said. It was a perception that was rooted in pre-sensation. Phenomenological philosophy refers to this as *future intentionality*, the idea that consciousness is directed not only toward the past or the present but also toward what lies ahead.

Researchers in psychology and neuroscience propose that the perception of time and of the self is profoundly modified during altered states of consciousness. A person might barely register time and then experience it retrospectively as compressed. Meditation practices also indicate a significant relationship between the perception of time and the perception of self: when the sense of self decreases, the sense of time decreases as well.

This might be the case when Adrian says, "I think I was in something that prepares time. Not like a clock. But like a movement that is already breathing you." In this context, the so-called *block universe theory*, based on the theory of relativity, suggests that space and time are merged into a unified four-dimensional space-time framework. From this follows the idea that past, present, and future are equally real.

Consciousness might not be extinguished when the brain goes offline; instead, it is liberated. What we usually call awareness is not generated but filtered. In this view, the brain does not act as a source but more like a tuning device. This model is gaining ground in consciousness studies. If there is such a field of awareness, independent of sensory input and neural architecture, then experiences like Adrian's may not be anomalies. They may be brief glimpses into the default state of unbound, undivided, and present consciousness.

Lev

The Child Who Spoke Without a Self

Lev had a severe allergic reaction that left him clinically dead for two minutes. He was only ten years old. Resuscitation was successful—but when he returned, he was different. He was not confused, shaken, or exhausted but quiet, alert, and almost composed. His parents expected a child in need of recovery; however, Lev did not seem weakened. Rather, he appeared permeable, as if something within him had opened and had never quite closed again.

When asked what he had experienced, he simply said:

"I was nobody. And that was good."

His parents initially suspected a defense mechanism, perhaps an early sign of dissociation. However, Lev was not sad or withdrawn. He was clear.

"There was no self anymore," he said. "But I was still there."

He did not describe a place, a path, or an event. Only a state, an existence without form, without history, and without voice. He did not experience himself as a person but as space. There were no thoughts or words, just perception—attentive, quiet, and completely present.

"I wasn't the voice," he once said. "I was what listens."

In the days that followed, Lev spoke in a way that sounded like the echo of something deeper than communication.

He said things like, "You can still exist, even when no one thinks of you."

Or: "The self is not an answer, it's a question."

And: "I was empty but full of everything."

His perception had shifted. Colors seemed softer. Time stretched. Sounds arrived like questions. He no longer felt the urge to explain or to be right. He simply watched. He simply was.

His teachers described him as "almost unusually quiet."

His mother said, "It's as if something inside him made space for everything else."

Lev grew up quietly. He studied philosophy. He asked questions without insisting on answers. He rarely spoke about what had happened. Only once, at the age of nineteen, he wrote in a notebook:

"Perhaps the greatest experience is not what you live through, but what makes you silent because you disappear into it without being lost."

Lev did not seek enlightenment or transcendence. He did not reject the world, but something in him had shifted direction. He no longer looked for meaning in things themselves but in the space between them. To him, identity was permeable. When asked what he believed in, he replied:

"I don't think we are what we think. We are what listens. What remains when everything else has passed through."

His life was quiet and uneventful on the surface. Those who met him remembered the stillness in his words, the precision in his gaze, and the way people sometimes felt a little more at peace in his presence, without knowing why. Lev had not been searching for something, but perhaps he had passed through something to return, quieter than before.

Scientific Commentary

Lev's experience brought back a shift in structure. His sense of self did not fracture; it merely stepped aside. In psychological terms, this was a gentle dissolution of the narrative identity, without losing the inner coherence. His words suggest a state of non-dual awareness, where the self is no longer the center but a quiet field in which thoughts arise and pass.

Neuroscience struggles to place such accounts. They are neither hallucinations nor symptoms of trauma. They point to a transformation of consciousness itself, shifting from actor to witness, from subject to space. Philosopher and consciousness researcher Thomas Metzinger maintains that human beings do not possess a "self" as a stable structure. Instead, he suggests that they inhabit a model of themselves that is continuously generated by the brain. People are not aware of the "self-model," so it appears entirely natural. In specific states, such as deep meditation, the "self-model" or "ego" becomes temporarily deactivated or transparent. However, consciousness remains.

In transpersonal psychology, with its focus on transcendent states of consciousness, the dissolution of the self can be explained through the approach of psychologist Abraham Maslow and his concept of peak experiences. During peak experiences, the sense of time and the sense of self or ego dissolve, and the person feels unusually connected to something much greater, for example, nature or meaningful interpersonal relationships. These experiences are accompanied by self-transcendence, which represents a further development of self-actualization, the highest level of human needs.

Self-transcendence is the highest form of human consciousness, in which altruism and universal values such as truth and goodness prevail. Therefore, peak experiences serve as moments of insight and clarity that imbue life with a profound sense of meaning. According to Maslow, peak experiences indicate that human experience also encompasses realms that go beyond the material or rational mind.

In Buddhist traditions as well, this concept has long been known as *anatta* or "the non-self," the understanding that a permanent, independent self does not exist.

Quantum physics, the science of the smallest constituents, has shaken our understanding of reality to its core. Atoms are not fixed particles, but rather fields of probability. Nothing exists statically. Everything hovers in states of possibility. Only observation brings reality into form. And what if consciousness is also not fixed? What if the self is embedded in a larger field that permeates us all?

Physicist David Bohm described the universe as a visible unfolding from a deeper, implicate order. The world we see—the material world—is only surface, and it is not connected to space and time. All information of what we are lies beneath the visible and is connected and interwoven. In this view, the self is not a sender, but a receiver and a tuning point. If consciousness is received, then perhaps death is not the end of a signal but the return into the broader field from which it came. Children who survive a near-death experience often speak of exactly this: a state, a connection. Not of death, but of something like memory. And they speak of something that had never vanished and had always been there.

REFLECTIONS

What happens when a child comes back from the edge and says that there was nothing? No place. No others. Only presence. We have found that not all children speak of light. Some children recall sound, motion, or stillness. And a few return with something that has no name. No place. No image. Only a knowing that has fallen silent.

It is tempting to dismiss such experiences as too abstract, too vague. Perhaps they are not vague but only precise in a different way. Conceivably, what Milos accessed was the space where everything ceases to be separate. The moment where direction dissolves. Where thought falls away. Where the mind no longer projects but receives.

Children like Milos are not confused. They are coherent in a way that does not require sequence. They remember something more basic than memory. A presence before perspective. A kind of awareness that feels like stillness that is not empty. It is full of something that cannot be seen but only known. What if this is not the absence of life, but its ground? What if this moment in between—this held breath of the universe—is not a pause between heartbeats but the place where the beating begins?

Milos returned with a calmness that did not need words. Perhaps children like him are not altered. Perhaps they are momentarily awakened into a dimension our language cannot reach. Maybe if we grow quiet enough, we might also hear it: a kind of thinking that no longer thinks. A presence that simply is.

What if time does not pass but prepares itself? What if each moment does not disappear but arrives and has been breathing toward us all along? Adrian was also in a place without light. And yet, everything was filled with warmth. Not outside, not inside, but between the breaths. There, where life does not speak but knows. Perhaps consciousness is not a flow but a field, waiting for us to pass through it, and when we do, it recognizes us. A place with no form, no direction, and no image but with a presence that breathes when we become still. Adrian was not alone. And yet, no one was there. Only a knowing that had always known him. A gaze that had long since seen him. And perhaps, when we ourselves arrive there one day, we will not need to remember. We will simply know that it knew us all along.

Adrian does not offer proof of life after death. He points to a reality larger than our comprehension, a reality in which consciousness itself becomes a permeable and resonant space, beyond past and future. Adrian's sentence remains striking: "I wasn't dead. But I was somewhere else." It points to a self that does not think but is felt by something greater than the individual ego. A self that is not an object but a relation.

Lev's story is a trace of what remains when the voice falls silent and only listening continues. Therefore, such experiences are difficult to put into words because every word would already be too much. He did not speak in symbols. He did not try to teach. He only said: "I wasn't the voice. I was what listens." A sentence that dismantles the self and repositions it as openness. In this also lies a paradox that has touched mystics and philosophers for centuries: that consciousness may not be what thinks but what receives, not substance but rhythm, not story but the silence beneath it.

Lev did not return to describe what he had seen. He returned to embody what he had become: quieter, gentler, less certain, and at the same time more present.

Some experiences do not instruct. They explain nothing. They simply shift the axis. Suddenly, it is no longer what is said that matters but what listens. Lev was simply different. Perhaps that is his gift: not an answer but a listening.

Some children return with messages. Lev returned with silence—a silence that demands nothing and yet sustains.

The Return That Is Chosen

Following a near-death experience, some children do not simply return—they *choose* to return. This decision does not arise through language or conscious thought but from a place of deep, inner certainty. In this context, it seems as if the choice to stay or to return belonged to them.

Although the space ahead feels both familiar and profoundly inviting, the return appears like an inner impulse toward the life that they left behind or a quiet knowing that something unfinished was left there. A moment in which the heart decides before the mind begins to understand. Many children report a guiding presence and a kind of boundary or threshold where they sense that going any further would mean that they cannot return.

These experiences offer a particularly precious glimpse into what reveals itself at the threshold of consciousness: that one does not always have to go on. When one returns, it is not only life that remains but also something deeper, something that reminds us of who we truly are.

Meera

The Voice Within

During a family journey in the Himalayas, eight-year-old Meera fell into a glacial basin. She had been walking toward a temple when she slipped on a wet stone and fell several meters into the ice-cold water. She was pulled out quickly, but she was not breathing. A doctor, by chance nearby, began resuscitation immediately, and Meera came back after several minutes. What she later told her parents left them speechless.

She did not describe a tunnel, and she did not float or fly. Instead, she expanded. Suddenly, she was no longer in her body but *everywhere*—in the trees, in the water, in the air, and in a space of light. This light was neither white nor blinding. It was soft and like the spectrum of all colors at once. The spectrum glowed from within as if it were something in her memory.

Then she heard a voice, but not with her ears and not in words. She heard it with her whole being. The message was simple and clear: *You can stay. Or you can go back. Both are good.* Meera did not answer. Instead, she thought of her sister. The voice replied: *Then go back. But carry the light with you.*

When she awoke, she was calm and clear-headed. She looked into her mother's eyes and whispered:

"I was in the light. And it knows who I am."

In the weeks that followed, Meera began to draw, namely spiral patterns, gates of light, geometric fields, and moving forms. These were things she had never seen or learned. She did not talk about it a lot but when asked, she would simply say:

"I don't remember with my head. I remember with my inside."

Though Meera came from a spiritual home, her experience did not reflect the traditional afterlife images of her family. There was not a god, judgment, or reincarnation. She only spoke of light, spaciousness, freedom, and knowing. Her parents, who were both rational in thought, could not explain why their daughter returned with this quiet knowing.

Scientific Commentary

Classic perceptual psychology ties hearing to vibration, stimulus, and neurons that decode waves. In this model, language is structured sound that is shaped into meaning. However, the voice that Meera heard was different. It did not have a tone or an origin.

Mystical traditions and contemplative philosophies, such as Zen, have long spoken of such perception or inner knowing. Some religious groups refer to this form of inner knowing as the presence of the Divine. The Quakers call it the Inner Light. Sufi mystics speak of the silent Friend. Psychologist Carl Jung described the Self as something that guides not through commands but through resonance. It is not speech that comes from the outside. It is knowing that arises from within.

Modern neuroscience has no full framework for this. Some scientists call it *intuition*, rapid, unconscious information processing that leads to a sense of certainty without being aware of what led to that conclusion.

Other scientists refer to the default mode network in the brain. This network is principally active if a person does not focus on the external world—for example, in resting states or daydreaming. It is active during self-reflection, thoughts about the past or future, and emotional evaluation. This means that the network is dedicated to inner perception and generates experiences that are not tied to external stimuli. In altered states of consciousness, such as near-death experiences, meditation, or trauma, this network changes its structure and enables experiences that appear like "answers without sound."

Amina

The Return Through the Light

Amina was seven years old when she was swept away by a flash flood during the season of the great rains. The earth had been dry, cracked, and still hesitant until it gave way. Within minutes, the dormant riverbed turned into a roaring current. The water did not come as a wave but as a silent, sudden, unstoppable flow.

Amina had been playing barefoot near the river's edge with her cousins. By the time the first scream rose, it was already too late. She cried out once and then she was gone.

Twenty minutes passed. When her body was pulled from the water, it was still and heavy. Amina had neither breath nor pulse. Her skin was cold and her lips were blue. The emergency team acted immediately with chest compressions, ventilation, and adrenaline. It was not a quick recovery. Hours later, long after the night had settled over the hospital, Amina opened her eyes.

She did not speak at first. Her mother sat beside her and was holding her hand. Amina's gaze was distant, aware but elsewhere. When she finally began to speak, it was not in fragments and not in fear. Her words came quietly, as if she had remembered something.

She spoke of being carried by the water that was not a violent force. She described it as a soft, invisible, and knowing "something" that did not pull her but held her. Then brightness came, and suddenly she found herself at the edge of a bridge that was golden, luminous, and woven like light. Its surface shimmered in strands that were threaded, as if made of movement itself.

Beyond the bridge lay a garden. That garden was not made of flowers in the usual sense but of voices. The petals moved like tone. Leaves shimmered in colors that felt like memory. The air itself carried a kind of music that was not a melody. Amina described it later as if color became sound, sound became feeling, and feeling became remembrance. She felt that the remembrance was about the home she had long forgotten.

Then, names began to echo softly, emerging from a distant place without clear direction. Amina did not recognize the names, but she felt as if they belonged to her; more precisely, she felt they were the names of people she had once loved and would meet again. Then she recognized a voice, although she did not hear it. The voice had no sound or gender, and it was clearer than language. It asked a single question, *Do you want to stay or return?*

Amina did not answer immediately, nor did she hesitate. She turned and there was her mother, who was crying. She stood behind a door that, as Amina later said, was not made of glass or wood but of wind and motion. Amina could not see through the door. However, she could feel how deeply her mother cried. The grief seemed to be like a field of tenderness.

In that moment, she made her decision.

"I chose to return," she later whispered. "The bridge folded like a wing."

She remembered that image with unusual clarity. She felt as if the light itself had become a gentle gesture that carried her and then became silence. Moments later, she opened her eyes.

After her return, Amina changed. The memory of the golden bridge and the garden of voices did not fade as a one-time event but stayed with her as an inner knowing that resurfaced throughout her life. She spoke less about her experience, but it was as if a part of her was always still present in that other realm. Her perspective on life and death was now shaped by a quiet clarity, something that set her apart from other children.

Amina developed a deep trust in the flow of life and in the connection between all things, including what had been and what was yet to come. In moments of stillness, especially when observing beauty or sorrow, she would sense that she was connected to something greater. She no longer heard the voices, but the knowledge remained. The act of choosing she had made during her experience was not just a memory; it was a lifelong decision, one that she consciously carried with her in every moment of her life.

Scientific Commentary

From a scientific perspective, Amina's experience opens new terrain. She reports contact with an intelligent, living environment. A reality in which sound and color are inseparable. A state where perception becomes presence, and presence becomes awareness. What appears to be a place is, more precisely, a form of consciousness.

Similar states have been reported in mystical experiences, deep meditative states, or other expanded modes of awareness. Vast spaces that are infused with color or permeated by light. Spaces that feel like fields of vibration or resonance. Sometimes, forms appear without fixed contours and become "fluid." But in a seven-year-old child with no spiritual exposure, no conceptual language for what she describes, the question becomes different. Where does such a world originate?

Adults who undergo near-death experiences often describe a "decision point of return," which typically occurs after the life review and after an encounter with a being of light or deceased relatives. In many cases, the reason for the choice is related to unfinished tasks or family connections.

In contrast, Amina describes a world in which memory is not bound to biography but to an inner knowing. A world in which sound becomes feeling, feeling becomes meaning, and meaning becomes the return.

Lenn

The Body That No Longer Hurt

Ten-year-old Lenn had lived his entire life inside a body that never felt quite safe. Born with a complex congenital defect, he had undergone multiple surgeries, been monitored by machines, and learned early that even the simplest movements could bring exhaustion. His body was never a companion. It was a fragile, unpredictable territory to manage, always one breath away from collapse. For Lenn, being alive meant being careful, vigilant, and in pain.

One morning at school, he collapsed without any warning, the result of sudden cardiac arrest. When the emergency responders arrived, he had no pulse. He was resuscitated and rushed to the hospital. For minutes, he was not there in the medical sense.

When Lenn regained consciousness, he was calm, rather than disoriented or afraid. He looked at his mother and said softly:

"It was the first time I didn't have a body," he spoke with astonishment. "It wasn't like flying," he explained later. "It was more like… nothing had to be held together anymore. I was just there."

He described a space filled with golden light that was warm, soft, and without borders. There was no floor, sky, or sense of time. There was only vastness and ease.

"I wasn't heavy," he said. "And I wasn't sick."

He did not meet anyone. There was not a figure or a voice. However, he felt a presence without form, a kind of silent invitation, and a question not asked in words but understood in feeling: *Do you want to stay with your body?* Lenn did not answer with language, but something in him said yes.

"Not loudly," he added. "Just gently."

When he returned, the condition remained, and his heart still struggled. However, his relationship to his body changed so that he no longer saw it as an obstacle. He did not speak of his heart with fear anymore but with calmness.

"It still skips sometimes," he said. "But now I know why I'm staying."

He still moved slowly, but there was no resistance in his gestures. Before the near-death experience, every step had been a negotiation. Now it was simply movement. Before the incident, his breath had been something to manage. Now it was something he noticed, like a quiet rhythm that belonged to him.

Scientific Commentary

Some near-death experiences shift something within: the way a child relates to their own body. Lenn's account belongs to this rare but deeply revealing category. What makes his story so significant, both personally and scientifically, is what he no longer felt: resistance, pain, and mistrust toward his physical self.

Children with severe, chronic diseases often live in a state of physical alertness. Medical interventions, recurring pain, and restrictions in movement shape not only their daily lives but also their inner experience. The relationship to their own body often becomes fragile. In psychology, this is called a disturbed *body-self relationship*.

The body no longer appears as a home but as a burden, something to be monitored, endured, and managed.

Lenn also lived within this tension. His body had never been a place of comfort, only one of exertion. Yet after his near-death experience, something fundamental shifted. He spoke of a space where nothing had to be held anymore. Where nothing was ill, not because the illness had vanished, but because it no longer defined him.

Such experiences challenge classic explanatory models. They cannot be easily classified neurologically or psychodynamically. In transpersonal psychology, however, these experiences are described as forms of *embodied transcendence*, states in which the self is no longer in resistance to the body but gently reconciles with it.

In many accounts, the process that leads to a return to the body is triggered by the prospect of reunification with loved ones in the "real" world. According to research, in adults, the return to the physical body is often accompanied by a kind of uncrossable border or limit, whereas in children, a return may occur without it. Only in a very few cases, as in Lenn's case, do children make a direct and voluntary decision regarding the return to the physical body. Adults are likely to consider the return involuntary, partly because they are sometimes accompanied by otherworldly guides as a positive presence. Children experience consciousness closer to a non-conceptual form, whereas adults' consciousness is shaped by language, biography, religion, and expectations. These factors also seem to influence the process of return.

Adelina

The Memory of Light

Adelina was eleven years old when she underwent emergency surgery. An undetected infection had spread suddenly and aggressively, with life-threatening results. The doctors fought for hours to save her. During the procedure, severe complications occurred: circulatory collapse, respiratory arrest, and critical instability. Externally, she appeared motionless. Internally, something was happening; something she would never forget.

She later recalled that everything first became quieter. There was a muffled but clear sound. Then she passed, as she described it, "through a membrane of stillness." She no longer felt like a body but like a single pure and silent point of awareness. She did not move, and yet she was moving. Then, the light came.

She described it as a field of warmth that opened like breath.

"It didn't blind me," she said. "It welcomed me."

There was no voice or calling. But she knew that she had been seen, recognized, and embraced. Then came a moment of knowing, without a thought or a vision. It was like a touch to the soul. She understood:

"This is not the end. It is only a passage. And I have a choice."

She remembered her mother, including a promise she had made to her. With a single inner impulse, she returned.

"It wasn't difficult. It was only sad. Because the light was so much more beautiful."

When she awoke, she cried quietly. She did not do this because of pain but from longing.

Adelina did not speak of her experience at first. Later, she began to talk to her grandmother, then to a psychologist, and finally to her mother. She told everyone the same story. She did not speak as if she were traumatized, but as if she were transformed. She became quieter and more transparent but not withdrawn. She listened to the way others breathed. She answered before questions were spoken. She dreamed of the light.

One day, while gazing at the sky, she said:

"It's always there. But we forget. Until it welcomes us again."

Scientific Commentary

Experiences like Adelina's are receiving increasing attention in near-death research. They are often described as forms of non-dual awareness: states in which the self dissolves without disappearing. Subject and object merge, time and space lose their boundaries, and a consciousness emerges that is vast. A knowing without words. A being without direction. A light that does not illuminate but receives.

The moment of decision does not follow an external logic. No dialogue, trial, or assignment. Just an impulse and a movement in the depths of one's being. Adelina remembered a promise and she returned, not as a resistance against the light but out of love for the world. In depth psychology, this is described as *implicit meaning-making*, meaning that is derived from inference and from automatic background processes.

Mystical traditions are familiar with such turning points: the contemplative stillness in Zen or the sacred silence in apophatic theology. *Contemplative stillness* refers to a state of awake, non-conceptual presence in which the mind settles into clarity without effort. *Sacred silence* describes a wordless, image-free form of encountering the Divine, in which conceptual thought vanishes and an immediate, non-dual awareness of presence emerges. In other words, it is not about knowledge but about presence. And perhaps Adelina's experience is precisely that: life's continuation on another frequency.

At the intersection of consciousness research and quantum physics, models are emerging that understand consciousness not as a product of the brain but as a nonlocal and embedded field. Concepts such as integrated information, nonlocal coherence, or morphic resonance suggest that consciousness is not produced but uncovered or revealed. From this perspective, Adelina's experience would be a moment of contact with a deeper order.

REFLECTIONS

In these stories the way of returning appears like a quiet offering, an inner recognition that life is still calling while the light is allowed to remain close. What unites these experiences is the moment when someone makes a choice. These children were not forced to return but presented with the option.

Meera's experience contains one of the most subtle but profound elements seen in many near-death narratives: the non-forced moment of decision. The voice she perceives does not persuade, warn, or direct. It simply opens two doors and acknowledges both as equally valid. The thought of her sister leads to Meera's return and replaces her answer. Regarding the voice, Meera's experience also points to a different form of knowing, one that is beyond sensory input and beyond the confines of speech—a message without medium. It is an internal resonance that does not need words to be understood.

This knowing is not something that can be simply explained by modern neuroscience. It transcends the material and enters the realm of the intangible, the space where consciousness meets the universe on a deeper level. Meera's experience is an example of this transcendence, a form of knowing that exists beyond the traditional senses, in the realm of pure awareness.

Amina's experience challenges us to think differently about the boundaries of consciousness. How can someone so young make a choice that reaches to the very core of life? What allows a child to access such memories, untouched by preconceived notions or spiritual frameworks? Where do these experiences originate, and how can they be explained? What she describes resists ordinary language because it belongs to a register beyond language. She does not speak of a place but of a state, a woven structure of meaning that is felt rather than seen. Where others might say, "I was guided," Amina says, "I chose." She speaks of conscious decision, not being sent back but choosing to return.

Her story contains a quiet turning point. A folding. A shift. A silence. She feels her mother is there. What if a child, beyond time, fear, or concepts, returns out of free will to preserve something? Amina's experience raises questions about the beyond but also about the here. Sometimes, an entire life begins with precisely this moment: The moment in which children remember why they have returned.

Lenn's experience marked a subtle and lasting shift in how he related to his own body, both physically and deeply spiritually. After his near-death experience, he no longer saw his body as something to fear or to manage but as a space he could inhabit without resistance. His disease did not disappear, but it no longer stood between him and himself. He began to live differently, not because the symptoms changed, but because his stance toward them softened.

For a child who had learned to live with caution, who had always watched his body like one watches a fragile bridge, the greatest shift was this: He did not feel the body as a burden anymore. Instead, he moved with it. He breathed with it. He rested inside it. His vulnerability remained, but it no longer defined him. He simply lived in it, as if the body had become a companion at last. And in that acceptance, something loosened in the space between pain and peace. He described it as sensation. He did not say "I saw." He said: "I wasn't heavy." He did not say "I was saved." He said: "Now I know why I'm staying." These are states and realignments. And perhaps that is what matters most: not that pain disappears, but that it no longer defines who we are.

Adelina's experience was marked by stillness and light. The light she described was warm, gentle, and welcoming. It was a subtle presence that enveloped her without forcing an explanation.

In her experience, the light was more a field than a place. A realm of potential, a space of pure being that did not require understanding, only presence. The light itself had recognized her and granted her the choice to return. Her return was not an act of will in the classical sense but a resonance. Something within her might have responded to the thought of her mother. Her return was a continuation of something deep, something she would carry with her: the light.

Perhaps this is the deepest message: That the light demands nothing and yet holds everything. That it does not call but recognizes. That it does not wait for us but lives within us in every pause, in every presence, and in a breath that deepens for no other reason than existence itself. And it remains until we are ready to be reminded of who we have always been.

Precognition and Transcendent Knowing

Time and again, children who have had near-death experiences report impressions that cannot be linked to concrete memories. They speak of something that is "approaching" or "in the process of becoming." In some cases, they sense something "already there." These are not predictions in the classical sense, no scenes or events with date and place. Rather, they describe atmospheres, inner images, and qualitative changes they have sensed with clarity, yet cannot fully name.

Some children do not cross a threshold. They stand at its edge and listen. What they encounter is not a figure or a place but a density. A knowledge that is perceived. Sometimes they sense it long before anything happens in the outside world. Without fear. Without confusion but instead with a strange, unchildlike clarity.

This chapter does not ask what children have seen, but what they have felt, even before language arises. It is about a silent field, a pre-verbal awareness that cannot be explained but can be remembered. It is about children's descriptions of encounters with a knowledge that was not learned but had always been there.

Noah

The Child Who Came from the Future

Noah's heart suddenly stopped when he was nine years old. He collapsed without warning in the schoolyard. The cause remained unclear—maybe an undetected infection or a subtle electrical disturbance. At first, his teacher thought he had fainted. Then he had a seizure. The seriousness became apparent when the emergency doctor arrived. The resuscitation lasted eight minutes. Eight minutes of silence within the heart and of undersupplied oxygen to the brain. Eight minutes during which time seemed to stretch beyond comprehension.

Noah was stabilized and ventilated. He survived. When he began to speak again, he said things that no one could quite place.

"I wasn't gone," he said. "Just somewhere else."

He spoke of light, vastness, and a stillness that sounded like breathing. Then he said something his parents would never forget:

"I saw the world when it was sad. But then it became lighter. Very slowly. As if it had understood something."

Noah described neither events, names, images, nor disasters. When he spoke of time, it did not sound like a child speaking. It sounded like someone who had seen time from the outside, not as threat or prophecy, but as calm awareness.

"Something difficult will come," he once said. "But people will become softer. First sad. Then warm."

He spoke of how animals would be heard more, how Earth would no longer wish to be ill, and how people would begin to remember things they had never learned: a light not from outside, but from within.

His words came quietly, and he did not share them in a dramatic or urgent way. His mother began to take notes because she felt that something important had been spoken. Something that should not be lost. The notes were not intended to prove or to analyze anything.

Scientific Commentary

Noah's experience cannot be classified as a life review or as a classical near-death vision. He did not see any images, figures, or places, and yet he spoke of something that was happening, in the process of becoming. It was a quiet knowing, as if the world were changing within him, even before it changed outwardly.

From a neuropsychological standpoint, such states might be described as altered time perception. In extreme conditions, such as cardiac arrest, hypoxia, or profound anesthesia, memory loses its linear structure. Experiences may overlap, extend, or arise simultaneously. Noah did not describe simultaneity, but rather a preverbal sensing of a future that was unknown, yet tangible.

In transpersonal psychology, there are concepts of an inner knowledge that is not acquired but remembered, as if touching an archetypal field that lay beyond age, biography, or language. Mystical traditions are also familiar with such states: an awareness that is not directed backward but listens to what has not yet been spoken.

This challenges classical models of memory and time. Ordinarily, we assume that events occur, are retained, and later are retrieved.

What if some forms of knowing do not move backward but forward—not because the future is predetermined, but because it can be felt as a possibility?

Terms such as anticipation, intuition, or *déjà vu* prove insufficient here. Neuroscience often explains such phenomena through the brain's ability to recognize patterns and simulate probabilities. Yet many of these children, like Noah, do not possess prior experiences from which they could have extrapolated. They do not project; they remember something that has not yet happened.

Such accounts touch upon the borderland of consciousness research, depth psychology, and modern quantum physics. The latter includes models in which time exists not as a linear sequence but as a field of overlapping potentials. In such fields, past, present, and future may not exist chronologically but simultaneously as possibilities. Under extreme physiological or psychological stress, access to this field may be altered, felt as a kind of resonance.

Empirical proof for such models is still lacking. Yet in newer theories of quantum cognition and integrative models of consciousness, interest in such boundary experiences is growing. They are no longer seen merely as fringe phenomena but as indications of a deeper order. Thus, Noah's case becomes an opening, an invitation to question our habitual concepts of time, knowledge, and consciousness. And perhaps to consider that time does not always look backwards. Sometimes, it also comes to meet us.

Elias

The Voice in the Dream

Several weeks before the accident, fourth-grader Elias began to speak of a recurring dream. He did not seem frightened, and he did not look for attention.

"There's a place," he said, "where everything is bright. Not like a lamp or the sun. More like... soft light. Wide light. It's quiet, but it still sounds."

His parents listened. They nodded, smiled, and gently asked questions. Something in them stirred, quietly and imperceptibly. It was more a subtle unease, rather than fear. Perhaps it was the clarity in his voice, or the calmness with which he said it. Or perhaps it was the sentence he added at the end, almost in passing:

"I think it knows me. And I think it's waiting."

He described a voice that called to him, not sharply but like a melody underwater. He did not understand the words. He only felt them.

"Gentle," he said. "And strong. At the same time."

He never called it a dream. He simply shared what he remembered.

The night before the accident, Elias went to bed early and peacefully. Later that night, his parents heard him murmur something in a content and smiling way. It was not as if he were elsewhere, but as if he were already a little further along.

The next morning, while putting on his sweater, he paused. His voice was barely audible and he whispered:

"If I go today—I'll come back."

His parents smiled quietly again, trying to interpret the child's words as poetry, not prophecy. However, something inside them froze, which led to a moment of stillness and a weight without shape.

Then the accident came. And the sentence returned, irrevocably. Like an echo that had always been there and only became audible after the silence had arrived. A key you recognize only after the door has opened.

His twin brother, Michael, remembered it, too. A few days after the funeral, he sat by the window and talked quietly about a dream he had before the accident occurred:

"We were on a hill. It was bright. There was light. Elias walked into it. I stayed behind. But I knew I could find him again."

He did not say this to comfort himself. He described what he had seen and what he knew. His expression was marked by clarity and the absence of metaphorical language. Michael's dream returned again and again with steady stillness, without a variation. Such consistency suggests not fantasy but inner order. A repetition that orients, rather than distracts.

One could call it psychological preparation. The calm, the coherence, and the spiritual tone of his account point to something deeper: a quiet readiness. In this clarity, the dream became more than a dream. It became memory, not of something lost but of something opened. A threshold that had never truly closed.

Scientific Commentary

Elias's and Michael's experience stands as an example of a rare, cross-cultural phenomenon: the premonition of death in children. Such experiences, which are quiet, recurring, and often in dreams, appear as condensed perceptions. Impressions that do not name but touch. Voices that do not call but accompany. Places that do not show but receive. Children between five and ten years old often report premonitory dreams casually, without the intent to interpret them. They do not analyze. They remember. Their language is clear, as if they were speaking from another frame of reference.

In the scientific literature, this is referred to as *death premonition*, an intuitive anticipation of dying. Such phenomena are difficult to classify, yet when they occur, they carry a peculiar clarity. They feel quiet and coherent. Dreams in which farewells happen before someone departs. Sentences that only gain meaning in retrospect. An inner knowing that requires no explanation, only silent acknowledgment.

Elias did not speak of death or of fear. What he shared was an experience: a sound, a light, and a voice that were familiar. Something that did not want to be explained. Something that was already present. In consciousness research, this is called *transcendent foreknowledge,* an inner sensing that does not stem from external information but rises from within. Perhaps body, mind, and soul perceive subtle changes before they manifest. Perhaps some children briefly enter a space where time opens and where the future is not known but remembered.

Elias's last sentence: "If I go today—I'll come back," is more than an expression. It carries a direction and an inner impulse. A certainty that was felt. His twin brother, Michael, who later recalled the same dream, reinforces its authenticity. These repetitions are not constructions. They act like a quiet concord between two inner spaces.

We cannot measure such experiences but we can witness them. They are not proof, but they indicate a cyclical, nonlinear consciousness. Children like Elias just describe, and in that lies their strength.

In psychological literature, this phenomenon is referred to as *anticipatory awareness*, an inner knowing shortly before the passage. It often manifests in dreams, images, or sentences, sometimes only in a changed tone of voice. Instead of fear or resignation, these experiences show orientation and a quiet readiness. Pre-death dreams and visions are also documented in palliative care, mostly from adults. Dying patients often report dreams or visions of deceased relatives or of journeys to the "other side." Animals also appear in these experiences, such as wild animals, pets, or birds. Although patients in palliative care receive medication, caregivers frequently report that the individuals remain mentally lucid.

So far, neuroscience has no terms for such coherent dream contents. It interprets dreams as processes of emotional regulation, memory integration, or stress processing. As mentioned earlier in transpersonal psychology, such states are understood as nonlocal accesses to consciousness, an idea reflected in Rupert Sheldrake's concept of *morphic resonance*, in which fields of collective memory extend beyond the individual mind. Carl Jung spoke of *islands of peace* that refer to inner psychological states of calm, wholeness, and centeredness. Consciousness researchers Stuart Hameroff and Roger Penrose developed models suggesting that quantum-physical information processes may occur within neuronal microtubules and result in a deeper level of brain activity, beyond synaptic networks.

All of these approaches share the idea that time does not only unfold linearly but can bend or open under certain conditions, especially in children, who have not yet learned what is impossible.

Timo

The Sentence That Waited Seven Years

This account is not about an experience related to death but about a different kind of premonition. Five-year-old Timo suddenly stopped in the garden, looked up at the sky, turned around, and said:

"The tree will burn. But you'll be grown up when it happens."

His mother laughed, not at him, but out of discomfort. He said it without drama or fear. Then he turned back and went on playing, as if he had said nothing at all. They forgot the sentence. Almost. It lingered like an uninvited, quiet, subtle, and unreachable sound in the room.

They remembered it seven years later. A lightning bolt struck during a summer storm. It hit the old walnut tree in the garden, which was the tree Timo had always sat beneath. Timo, now twelve years old, stood at the window and saw the tree burning.

"I said it, didn't I?" he asked in an untroubled way. "Back then."

His mother had gone pale. She had not remembered it consciously, but it had remained.

"I don't know how I knew," Timo said. "But it was like a sentence that had always been there. It was just waiting for the moment."

He did not say it with pride.

"It wasn't knowledge," he said. "It was... remembering the future."

Scientific Commentary

Timo's experience eludes the classical understanding of time: How can a child remember something that has not yet happened? How can a sentence lie dormant and unnoticed for seven years and then unfold as if it had simply been waiting for its moment?

This points to a different form of remembering, one that does not operate chronologically but cyclically and multidimensionally. From a neuroscientific point of view, Timo's experience seems impossible. Memory is regarded as backward-looking. It stores what has been and recalls it later, linked to neural processes in the brain.

In physics there are models that describe precisely this: time as a field, not as a flow. The block universe theory, introduced earlier, assumes that past, present, and future exist simultaneously, like the pages of a book that is already complete. Even if events appear to us as a sequence, on a deeper level, they are interconnected.

Philosopher John William Dunne spoke as early as 1927 of *serial time*: the idea that deeper layers of consciousness may move freely between levels of time. In such layers, something might be perceived that eludes everyday thought, not as a prediction, but as a resonance.

From a psychological perspective, Timo's experience might be interpreted as *precognitive intuition*, an unconscious grasp of possible futures. However, his words go beyond typical intuition. They are neither tentative nor uncertain. They are received as a memory.

Marlon

The Bright Shadow

Marlon developed a habit when he was ten years old: He began turning around at the same time each evening. Not abruptly or anxiously, but rather as if someone were responding to a quiet call. He said he did not hear a voice or see a figure. He only felt a presence, like a soft, wordless knocking in the room. Marlon called it a "bright shadow," something that did not obscure but sharpened his perception, a kind of awareness.

"It's like a word hanging in the air," he once said, "but no one says it."

His parents asked if he was afraid. Marlon shook his head as he did not feel pursued. He felt accompanied.

"It's not someone," he explained. "It's more like a knowing that had already arrived."

A few weeks later, during a hospital stay, Marlon had a severe reaction to medication. His circulation collapsed. For several minutes, he was clinically unconscious. When he was stabilized, a nurse asked if he knew what had happened.

Marlon replied: "I already knew. It had been there. Every night."

He did not describe a classic near-death experience. What he spoke of was a clear, condensed, and calm state.

"I was a point. And the point was everywhere. And it knew I was coming." Not as a person or as a self, but as a movement through space.

Later, he drew a black rectangle with a glowing center. It resembled either an iris or a window deep in the dark, a shaft of light that did not dazzle but waited. When asked what it was, Marlon said:

"You go through it. But it's not a path. It's a question."

He insisted that he had not chosen to return.

"I just heard that it wasn't yet the moment to stay."

In the weeks that followed, Marlon began to speak more softly and more precisely. He began to use words that did not belong to a child's usual vocabulary—terms like "threshold," "line of sight," and "plane of stillness." He wrote them on scraps of paper, tucked them into books, or slid them under his pillow as markers.

His teacher noticed a change in class. He was focused and present in a quiet, listening way. When he spoke, the room paused. In one essay, he wrote:

"I think some things aren't meant to be seen. They're meant to be felt when no one is looking."

When his mother asked if he remembered what had happened, he said:

"I didn't notice it with my head. That's why I can't forget it."

For him, it was not a rupture but a part of a continuum that had already begun and did not end when he returned.

Today, many years later, Marlon rarely speaks about that time. But when he does, he remembers it with quiet certainty.

"I was never alone. Not before. Not during. Not after. Only the world was often too loud to hear it."

Scientific Commentary

Marlon's experience not only challenges the classical concepts of near-death experiences but also opens the possibility of a subtle form of inner perception. A sensing that does not respond to external stimuli but to something that announces itself beyond the visible.

The "bright shadow" that Marlon sensed each evening was neither a hallucination, a dream, nor a classic apparition. It was a presence, a tangible field in space, without form but with density. In transpersonal psychology and consciousness research, there are descriptions of such phenomena: a preverbal knowledge that is not learned but felt. This kind of perception is not cognitive but resonant. It precedes language and acts through the body. In early human development, learning takes place through sound, rhythm, touch, and emotional charge through resonance.

Resonance might also explain why people often feel a vague premonition without knowing what it refers to. In neuroscience, research suggests that human beings do not primarily react to events but rather anticipate them. This *predictive coding* of the brain compares expected input with the actual stimuli perceived by the body.

When Marlon said, "I was a point. And the point was everywhere," he described a state of self-awareness that was not fixed in space or identity, but was dispersed, fluid, and listening. Such descriptions align with concepts in quantum physics—for example, in the theory of the block universe mentioned earlier, according to which past, present, and future exist simultaneously. It also exists in concepts of coherent fields of consciousness, as formulated by physicists such as David Bohm or Federico Faggin: consciousness as a nonlocal field in which information is not transmitted linearly but felt through vibration and presence.

REFLECTIONS

What does it mean when a child speaks of something that has not yet happened and yet speaks as if that moment had already been lived? And this with a quiet, profound certainty. What is this? A coincidence? A feeling with a calendar entry? Or is it memory, not of what has passed, but of what is yet to come?

Philosophically, this question is an old one: Is there an inner time, a resonance that follows not clocks, but meaning? Thinkers from Augustine to Henri Bergson described time as a lived depth, as a qualitative duration rather than a measurable sequence.

These children do not prophesy. They do not warn. Yet at times, they stand in relation to something that is not yet formed and yet already perceptible. Noah did not speak of events but of a changed atmosphere. He did not describe what the world would do but how it might feel. Sadness softens. Grief turns to light. As if the world had remembered something it had long forgotten.

Perhaps children like Timo meet an inner pattern or an order that is not linear but cyclical. Within it, time does not appear as something that passes, but as something that reveals itself, like a circle where beginning and end quietly touch. Where some sentences wait to emerge, like an inner trace that had already been laid. And when they come, they do so not as an explanation but as an echo.

And sometimes, in such moments, time loses its direction. Timo stood still and spoke a sentence no one understood, until years later, when the tree burned. Then the sentence returned like a stone thrown in a circle, remembered only when it comes back. There are moments when what seems separate, particularly language and feeling, future and memory, begins to overlap. We may not experience everything that happens. There are sentences that rest within us until they awaken, because we are finally ready to hear them.

In a way, that was exactly what Elias made so clear. Perhaps he did not foresee his death but felt a call, from a space already familiar to him. In that sense, his last words would have been not a prediction but a gift to prepare and to ease the leaving.

Maybe the future is not only what lies ahead of us, but what already lives within us, waiting to be recognized. The words Marlon wrote down, placing them like markers in his room, were not messages but traces of an inner space he had entered. Perhaps it was not Marlon who recognized something but the field perceiving through him, resulting in the most radical form of remembrance: not the glance backward but forward. Remembering what has not yet been—children, the future, and the depth of time. Instead of referring to things that have not yet happened, the future may be what is already in the process of becoming within us.

When Time Stands Still

There are moments when time ceases to elapse, not in the sense of an end, but as a pause. A breath in the rhythm of existence. In such moments, consciousness loosens its command on sequence. This chapter invites us into one such space where awareness does not leave the body but expands beyond it. Where the boundary between two souls softens and something essential begins to speak.

For most of us, time is something that passes. A sequence of moments, structured by numbers, tracked by instruments, and fixed in calendars. What if time cannot be measured? What if it does not pass but sounds? What if it does not act upon us from the outside but pulses within us, as a field that unfolds through perception?

Time and death are deeply connected. When time ceases to elapse in the moment of death, for many people such moments serve as a gentle greeting, a symbolic gesture of farewell, or a subtle sign that our connection to those whom we love does not end with death.

Matthew

The Clock That Stopped

Matthew was nine years old when the unthinkable happened. His twin brother, Elijah, died in an accident. The two were inseparable. "Two halves of one whole," their parents would say. They did not just share a room. They shared their dreams and exchanged words that were incomprehensible to others. Moreover, they shared a closeness, as if fate had always known they belonged together. It simply existed.

That day, Matthew stayed home with a mild fever. He was just tired and quiet, nothing alarming. Elijah had gone to school alone. Late in the morning, Matthew lay unusually still in bed. "Too still," his mother would later recall. As if he was not quite there.

Then, around half past ten, he suddenly sat up, looked at the wall clock, and whispered: "It stopped." And it had. The clock's second hand had frozen mid-motion. When his mother entered the room, moments later, the clock resumed ticking, as if nothing had happened.

Later that day, the phone call came. Elijah had been struck by a car. Time of death: 10:27 a.m.—the very minute the clock had stopped.

In the days that followed, Matthew spoke little. He was not withdrawn but inward. As if he were listening to something the world could not hear. On the third night, just before sleep, he said quietly:

"He was there. In the doorway. He said goodbye."

He described the moment without fear. He had sensed a peaceful, familiar figure in light, not speaking but smiling. Then, as if carried by something unnamable, Elijah disappeared into "a sound that felt like warm wind." Matthew was sure: that was Elijah.

"Not like thinking," he said. "More like knowing."

Matthew changed after Elijah's death, though not in the expected way. He did not become withdrawn. Instead, he became spacious, as if something had opened inside him, like an open-minded horizon. He sometimes spoke of Elijah in ways that carried stillness.

"I dream of him," he said, "but it doesn't feel like dreaming." Or: "I hear him when I'm not thinking." And again: "He's not gone. Just somewhere else."

Scientific Commentary

The story of Matthew and Elijah lies at the subtle intersection of psychology, consciousness research, and intuitive perception. It does not fall within the classic spectrum of near-death experiences and yet it opens a neighboring space: that of inner synchrony. This phenomenon is frequently reported between individuals who share a deep emotional and relational bond and with striking regularity between twins. Individuals who synchronize their behavioral, emotional, or physiological states with each other over time.

Stories of perceptive connection between twins are not rare, especially in moments of sudden separation. Studies document numerous cases of what is sometimes called "twin intuition": one twin senses the pain, fear, or even death of the other, often over great distances and in the absence of any communication.

Psychology and neurobiology offer concepts such as mirror neurons, emotional resonance, and unconscious synchronization.

Shared prenatal development, identical genetics, and mirrored early experiences seem to create a synchrony that extends beyond normal relational empathy. These theories describe how deeply bonded individuals may sense each other's pain or absence. However, Matthew's experience reached beyond that framework. The clock that stopped. The appearance in the doorway. The wind that was a sound. These details do not fit into ordinary models of perception.

In research literature, such phenomena are sometimes called *death synchronizations*, moments in which clocks stop, animals become agitated, candles extinguish, lights flicker, or rooms shift their atmosphere at the exact time someone dies. These events cannot be proven, but they have been reported again and again, especially in stories involving twins, soulmates, or between parents and children. When two beings are bound beyond time, one thread sometimes trembles as the other is released.

In addition to an observable physical anomaly—the clock that stopped at the exact minute of Elijah's medically confirmed time of death—there was a nonverbal, visual encounter: the appearance of Elijah in light, peace, and stillness.

This phenomenon is called *after-death communication*, and it is characterized by spontaneity and by the living person's feeling of having direct contact with someone who has died. The relationship with the physically deceased person is often described as a vivid continuation. Researchers have discussed possibilities of explanation such as hallucinations or dreams. Hallucinations often result in feelings of detachment from oneself or from the environment, which are called *depersonalization* or *derealization*. But many aspects of after-death communication do not correspond to hallucinations—for example their beneficial effect, palpable reality, and persistent memory.

Oliver

The Clock That Ran Backwards

A clock also plays a role in the account of the twin brothers Oliver and Julian, who were seventeen years old when Julian unexpectedly died. The death happened in the middle of the night, while Julian was far from home. The cause was a ruptured aneurysm.

Just the day before, the twins had exchanged messages, as they always did. They were not just close. They were a "rhythm," two lives that moved in a parallel, intuitive, and inseparable way. They shared jokes, silences, and instincts.

Oliver was alone in his room that night, writing a message to Julian. His fingers paused above the keyboard. He glanced up, not knowing why. The wall clock across from him caught his attention. For a moment, it looked normal, and then it changed. The minute hand began to move slowly, not forward but backward. Tick by tick as time reversed.

At first, Oliver thought he was imagining it. Maybe it was a trick of tired eyes or a mechanical glitch. But the clock kept going relentlessly until it stopped at exactly 12:55 a.m. Ten minutes later, the phone rang. Julian was gone. The official time of death: 12:55 a.m.

In the days that followed, Oliver did not speak much. When he did, his words were imbued with substance.

"Something stopped," he said. "Not just the clock. Not just time. It was like my body stopped sending signals. I wasn't a sender anymore. I was a receiver."

That moment, he later said, did not mark the end of something. It marked the beginning of a different kind of perception. In the nights that followed, Julian began to appear in Oliver's dreams. Always in the same place, which featured a corridor of soft, glowing light, almost like mist. Julian never spoke but he always turned and smiled.

"He said goodbye," Oliver explained. "Not with words. With that look. And with the clock."

Since that night, Oliver has never looked at time the same way. For him, it is no longer a line but a field. A dimension that can widen, fold, or stand still.

"Sometimes," he says, "everything slows down. And sometimes it just stops so you can really see what's there."

The clock that stopped that night still hangs in his room. It works again. The hands move normally now. To Oliver, it is no longer merely a clock, but also a witness. A turning point and a quiet reminder that some threads do not unravel when cut. They stretch into places we do not yet have names for.

Scientific Commentary

One of the often-reported synchronistic events surrounding death is the stopping of clocks. Nevertheless, Oliver's story carries a rare quality: not the cessation of time but its reversal. The clock was later examined and there was neither a technical fault nor a misalignment. The mechanism was intact.

Some might say that Oliver, deeply bonded with his twin brother, picked up on subtle cues: nervous system resonance or a kind of cellular empathy. However, other voices in the field suggest something else: a connection that is not bound by space. A form of knowing that does not travel through time but exists beneath it as a presence. Sometimes something happens, and time slips with a softness that escapes the eye. A clock pauses. A breeze shifts, though no window is open. Someone, somewhere, senses a change, without knowing why.

These moments have a name in the literature of extraordinary experience: *synchronicities*, meaningful coincidences without causal connection, yet charged with inner resonance. The term was coined by Carl Jung to describe those events in which the inner and outer worlds coincide in a mysterious and meaningful way. They appear without pattern, without explanation, but they are felt. They are like a ripple moving through the invisible, linking two places, two people, or two states of being, just for an instant. For Jung, the psyche and the material world are not separate domains but two different manifestations of a deep underlying unity. He assumed that both psychological and material processes arise from a common ground; a single, undivided reality from which psyche and matter only differentiate at a subsequent stage.

Coincidences in the form of synchronicity seem to occur more frequently when universal and basic psychic patterns are activated in times of profound change, in dreams, crises, grief, or threshold experiences. Universal patterns are in the collective unconscious of all human beings.

The previously mentioned theory of the implicate order by quantum physicist David Bohm does not rule out such experiences but interprets them differently. In this view, time is not a flow but an expression of unfolding. What we experience is not all that exists.

Certain events, such as a clock running backwards, could be clues to this deeper structure: an order that sometimes reveals itself when the surface develops cracks.

Lilia

The Sentence That Was Never Spoken

Lilia lost her father when she was six years old. He died in the middle of the night, without any warning. Again, in this case, a ruptured aneurysm played a role. It was one of those moments that leaves no time for words but only silence. Her mother found him in the morning. The apartment was quiet. The coffee machine had not been switched on, and the newspaper lay unread and untouched. Only time had moved on.

Lilia was asleep in her small bed when something entered the room. That "something" did not arrive loudly or with footsteps. It was more like a subtle and hesitant feeling. Lilia did not wake immediately. It was as if something stood at the edge of awareness, waiting for its own moment. When her mother sat beside her a short while later, Lilia looked at her. No words had been spoken, but something had changed.

Later, she would say:

"I knew. Before anyone was sad."

In the hours that followed, the world moved more slowly. Words, gestures, and even light seemed to hesitate. Lilia did not speak much. She drew with wax crayons. She turned her stuffed animals over, one by one.

She watched the sunlight as it crept along the wooden floor. Her mother tried to explain gently and carefully what had happened to her father. However, it seemed as though Lilia had already stepped one pace further. She was not confused or overwhelmed, just deeply quiet.

Two days later, she said almost casually:

"I think he wanted to say something."

It was not a wish. It did not sound like comfort. It had the tone of certainty.

On the third night, she lay awake for a long time, or perhaps she was asleep, she could not quite tell. The room was completely quiet and dark. And yet, something changed within her.

"It wasn't like someone came through the door," she said. "It was more like… when you already know someone is about to call your name. And you already know what they're going to say."

She felt no fear or tension. Instead, she just felt warmth and a soft expansion in the air around her. Although she did not see a figure or hear a voice, she felt that something was there, a familiar presence that was barely separate from her own.

"He was there," she said later. "Not where I was looking, but where my heart was looking."

And then something happened. Lilia received a message that was not given through words or sound, but as if placed into her without language:

I'm not gone. And you don't have to carry everything alone.

It did not come as a thought. It was simply there, like a reply to a question she had never asked aloud.

In the morning, she got up, dressed herself and walked into the kitchen. Her mother looked at her for a long moment, in a quiet and searching manner. Lilia climbed onto the stool beside the counter and said softly:

"I think he said it," as she reached for a piece of bread.

Scientific Commentary

Lilia's experience relates to a phenomenon often described in the context of death, loss, and transition but which has rarely been systematically researched: the quiet, subtle, intangible sense of presence that nevertheless feels profoundly real. Children especially speak of such moments in a way that feels familiar and entirely natural, without drama or the need for explanation. They report them as an encounter without a voice, or a form, and yet full of meaning.

Such after-death communications are often documented. Studies suggest that up to 70 percent of grieving family members report some form of nonsensory perception following the death of a close loved one: a feeling of closeness, a familiar scent, a quiet inner phrase, the sudden knowledge that "someone is there." These impressions often occur in moments of stillness as something that briefly becomes perceptible, even when no expectation of making contact is at the forefront of one's mind.

Lilia did not receive a sentence in the usual sense. There was no voice or message from outside. Rather, it was an inner knowing that was not thought but felt. Therein lies the depth of such experiences: they bypass language. They reach a level on which communication is possible without words and yet with certainty.

Children seem particularly open to such perceptions. Perhaps because they have not yet fully internalized the strict separation between what is visible and invisible, between being here and being gone. Lilia's wording, "He was there. Not where I was looking, but where my heart was looking," points to a kind of seeing that lies beyond the outer senses. Consciousness research speaks in such cases of *non-sensory understanding*, a direct form of inner knowledge that is not derived from the five sensory impressions: sight, hearing, smell, taste, or touch.

Grief psychology also points out that children experience death differently. They do not frame their experiences within the concepts of ending and farewell. For them, connection is not a state dissolved by death but a field that continues to exist. In this field, a sentence that was never spoken can nonetheless be completed. And sometimes, that message alone is enough for life to continue.

Nia

The Sound of Time

Eight-year-old Nia had a near-death experience during a seizure or a neurological interruption. For Nia, it felt like a transition into sound.

"I heard time," she said.

It was not the ticking of a clock or the rhythmic hammering of a machine. It was something else entirely.

"It was like music, but without melody. Like a breath listening to itself."

She described being in a room that did not move, yet it was filled with waves.

"I could feel that everything was vibrating," she said. "Not fast. Not loud. But deep."

It was not an external sound she heard but an inner motion. A vibration that did not come from outside but from within or perhaps from everywhere. After that, a moment came in which everything grew silent.

"So silent," she said, "that I knew something was about to happen."

She did not fall, she sank and dissolved out of a body of light, as she called it.

"It felt like I had another body—and that body fell through me."

When she returned, she heard nothing except an inner echo. This echo was a lingering vibration without origin, or a quiet breath from another sphere. She drew spirals. She spoke words like "slower" or "farther" or "backward into forward."

And sometimes, she would close her eyes and whisper:

"It's still sounding."

Scientific Commentary

Nia's experience touches a dimension of near-death perception that classical medicine rarely considers: the auditory shape of time, the feeling that time does not pass or flows but sounds and vibrates. In modern physics, time is not a fixed stream. Albert Einstein spoke of the "illusion of time." In his theory of relativity, simultaneity depends on the observer. Furthermore, time is malleable; it can stretch or contract, and it is connected with space.

Quantum physics goes even further: it describes states of superposition in which past and future are not clearly distinct, but instead overlap. Time does not appear as an arrow but acts in a more elastic, rhythmic, and nonlinear way. Time is not divided into sequential fragments but exists as a field of potential, comparable to a piece of music that already exists before it is played. Nia's experience seems to have touched precisely that conception of time. She describes time as a continuum that is revealed to her like an inner sound, a space of consciousness she can feel.

Some theories even suggest that time is not fundamental and that, at the most basic level of physical reality, it plays no role at all. Physicist Carlo Rovelli's theory of the *relational quantum mechanics* suggests that motion, change, and causality emerge from the relations between physical systems. Time arises in the act of these relations.

In many mystical traditions, time is not seen as a linear flow but as a pulsating presence. Philosopher Henri Bergson distinguished between *temps*—measurable, spatialized time—and *durée,* experienced duration: that inner time which cannot be grasped in seconds but appears in consciousness as stretched, condensed, or dissolved.

Transpersonal psychology also knows of such experiences. The self is not understood as a fixed ego or identity but as a transition in the field of awareness. The "body of light" out of which Nia dissolved echoes states of deep meditation, ecstasy, and out-of-body awareness described in many spiritual traditions. Such images often appear in children but rarely with the clarity with which Nia spoke.

Contemporary consciousness research increasingly recognizes that time is not an external metric but a psychophysical phenomenon. Studies show that our sense of time is closely tied to our state of attention, emotion, and bodily awareness. In ecstatic or traumatic moments, time expands. In meditative states, it dissolves. Neuroscientist Wolf Singer describes time as a function of *synchronized neuronal oscillations.* However, these rhythms are not beats. They are patterns. They do not structure time itself but the experience of presence.

When Nia says, "It's still sounding," she might speak of a depth that has not passed but continues to move through her. In spiritual teachings, this state is known as *kairos*, a deep, qualitative, timeless moment in which time unfolds outside the linear sequence, outside the usual chronological flow of time. In Eastern philosophy, it echoes the notion of *satori*, a sudden moment of silent clarity beyond thought. In Hebrew, the word *kavanah* describes an inner alignment within a moment of intentional, quiet awareness.

REFLECTIONS

Children rarely speak in straight lines. They begin in the middle or everywhere at once. What they describe often feels less like something that happened and more like something that is still happening. Their words do not follow a linear sequence but a circular movement. Time does not pass them by; it flows through them. When adults ask children about their near-death experiences, for example, "But when exactly did it happen?", the child just looks back quietly, as if to say that the question no longer applies here. This raises a quiet but profound question: What if time is not a line we follow but a field we enter, especially when someone crosses its edge?

We think of time as measurable, as something moving forward in fixed steps, marked by clocks and calendars. What if time is not a thread outside of us, but a spiral unfolding within us? What if we do not travel through time but expand in it, moment by moment, memory by memory? Some children speak of things that happen later but remember them beforehand. They tell of voices that spoke before events occurred. Of images that appeared before they took form. The present becomes layered or multi-layered. Moments no longer follow one another; instead, they rest within each other, like overlapping footprints in the sand.

Neuroscience often says the brain constructs continuity; that it fills gaps and arranges fragments. Maybe the brain does not construct meaning. It might carry it like a vessel carries water. And perhaps, in certain states of consciousness, this vessel opens, revealing a field and a depth. Perhaps time is not what passes us by. It is we who move within it. And sometimes for one breathless moment, we pause. Not outwardly but inwardly. And in that stillness, we hear something, not with our ears but with the part of us that lives beneath time. A pulse older than language. A sound made not of tones but of meaning remembered.

Lilia's and Matthew's experiences speak to the phenomenon of after-death communications. These moments are neither hallucinations nor dreams but a form of knowing or a shift in orientation, experienced in a wordless, profound, and deeply real manner. Lilia and Matthew were quietly invited into the liminal space where language dissolves and connection remains. Perhaps that is also a kind of passage, not for the one who goes but for the one who stays. A silent crossing marked by attunement and presence. And maybe, in that moment when one life leaves, another begins to perceive, with something that listens when the world falls quiet and that knows without knowing why.

After-death communications are integral, complete, and often difficult to put into words. For those affected, it feels more real than many things seen with the eyes or heard with the ears. Such moments are no less true because they are subjective. They are often significant for that very reason, because they touch the subject in its depth.

Some bonds are not learned. They are simply there, from the first breath, from before language. Sometimes even from before birth. Twins often speak of a closeness that has no distance, no effort, no explanation. It simply is as if consciousness could stretch across two bodies, like breathing, sensing, remembering together. Some connections do not end when the body does. They do not need to continue because they never began. They were always there. And in the moment of parting, they become visible. Not as appearance or light, but as knowing, as warmth. A sudden stillness in the middle of the day.

Perhaps dying is not the end but the moment we briefly wake up to remember that we were never truly apart. There may be even regions of consciousness where we are connected long before we ever meet. Where knowing is not passed between us but shared by virtue of being alive together. Where presence resonates.

Oliver did not "see" anything in the traditional sense. No being called to him. And yet he knew something was near. That knowledge was not predictive. It was not visual, cognitive, or symbolic. But it was unmistakably present. In the framework of transpersonal psychology, such a state is viewed as an opening. A moment when consciousness does not act but receives. As Oliver himself put it: "I wasn't a sender anymore. I was a receiver." What Oliver described was not imagination or belief. It was unspoken, immediate perception, without any need for justification. In the moment his brother died, something in him shifted, from sender to receiver. As if the signal came from a space they had always shared.

From a philosophical perspective, deeper questions emerge: Is consciousness inherently individual, or can two inner worlds converge, temporarily forming one shared field of experience? In the case of twins, the sense of separation often dissolves. They do not experience the other as "other." They feel as if they were living across two bodies. In Matthew's words, it was not a loss but a shift, a transformation. And perhaps this is what makes these accounts so deeply human, not only for research, but for all those who have ever sensed that love, once formed, does not vanish. It changes its form. It becomes presence without body, thought without words, and stillness that remains, even when one breath stops.

Nia's experience also opens a space for another understanding of time. She did not hear a sound. Through that silent inner listening, she felt vibration, an inner resonance that encompassed everything. Perhaps it was the sound of time. Perhaps she herself was the one who, for one brief moment, became entirely still. And, by becoming still, she heard everything. When Nia says, "It's still sounding," perhaps she speaks of a depth that continues to move through her. A resonance, a reverberation not based on memory but on the present moment.

These stories raise more than our emotions. They open questions. Can time be sensed, not as sequence but as substance? Can closeness shape reality, not just memory? And what if there is something in us that does not experience loss as separation but as transformation, not as the end of connection but as the change in its direction? Perhaps time is not a continuum. Conceivably it is a field and is audible when everything else falls silent. And the moment in which it sounds is not a passage but an awakening at the very heart of consciousness. No before, no after, only the vibrating now.

Memory of Life

Many adults have reported that their near-death experiences involved seeing a passage through a tunnel and an encounter with a light, often accompanied by a life review. Although tunnels and lights are sometimes part of the near-death experiences in people of any age, younger children rarely speak of having a life review.

Individuals who return from the threshold of death often come back changed and adjust the way they lead their lives. The experience itself imprints deeply on their memory and accompanies them as they develop a new appreciation for life. Relationships gain importance, while material concerns lose their weight. Some people describe an inner knowing that cannot be put into words, along with a sense of connection that reaches far beyond everyday awareness. Their way of dealing with stress, illness, and loss often shifts toward a calmer, more grounded attitude. For many, the near-death experience becomes an inner compass that shapes their understanding of meaning, time, and consciousness in the long term.

In addition to near-death experiences in which people see scenes from their own lives, there are accounts of images, memories, or encounters that appear to originate from another time, another place, or even another life. These impressions often do not have a recognizable connection to the individual's personal history. Such experiences lead to questions that have fascinated humanity for centuries: What is memory? Where does identity begin? And can consciousness access or retrieve information that does not belong to one's own life?

Anna

The Tunnel After the Impact

Anna was fifteen years old when she was hit head-on by a car while riding her bicycle. The impact hurled her several meters through the air. She sustained multiple fractures, severe lung contusion, internal bleeding, and a brief cardiac arrest. The situation was life-threatening. When she later regained consciousness, she did not speak of pain or of the accident, but of a tunnel.

She described how she suddenly found herself floating above everything, without knowing how or why. She saw herself lying below, motionless, surrounded by blue lights and unfamiliar voices. Yet she felt nothing. There was no pain, no fear, but only stillness and light. Then the space narrowed. A tunnel opened that was not made of stone but of flowing brightness. It was wide yet directed. She was not pulled, and yet she moved, not by choice; instead, through a natural motion of being.

At the end of the tunnel, there was no place, no person, but something alive, a state. A kind of knowing that required no questions, because all answers were already there.

"I didn't see," she said. "But I knew. I didn't ask. But everything was clear."

As she approached the light, she relived her life. This happened not as a linear sequence, but rather as an immediate sensation. Every encounter, every smile, and every pain, all present in fractions of a second, and yet completely felt. She did not just recall what she had done. She felt how it had touched others.

"I was me," she later said. "But I was also the others."

Then came a moment of decision. There was no voice, no command, only a quiet, undeniable question: to stay or to return. She felt her mother's body, which was broken and weeping, like a memory reaching across space and time. And with one last glance into the light, she made her choice. She returned.

When she awoke in the intensive care unit, she cried, not from pain but from loss.

"It felt like I had left the real place. This was smaller. Harder. Narrower."

She brought something back: a new stillness and a different way of seeing. Her parents described her as changed. She spoke more quietly, smiled differently, seemed more focused and more transparent. She began to explore meditation, ethics, and consciousness. She read about quantum physics and spiritual philosophy, became socially involved, and spoke to other young people about vulnerability, not with drama but with clarity.

Her teachers also noticed the change as she expressed a new depth of thought. Her friends spoke of a different kind of presence. She herself put it this way:

"I know now that nothing is lost. Not a smile. Not fear. Not love. Everything continues. Even when we can't see it anymore."

Years later, Anna accompanies others. She works with teenagers who have experienced deep trauma and helps them find words for what often resists language. And sometimes, when the room grows very still, she softly says:

"I know the light isn't gone. It's just a little further back. But it's there."

Scientific Commentary

A life review in near-death experiences is reported more often by adults and adolescents. One possible reason is that young children have less autobiographical material to draw on. In these reports, the life review is often not a "film" that plays back but a moral-empathic space. The retrospective review of one's life does not follow a chronological order, and it does not appear selective or distorted. Everything appears simultaneously, so time becomes depth rather than sequence.

According to the Professor of Health Science and near-death researcher William J. Serdahely, the lack of life reviews in younger children may also reflect that children might not yet be psychologically capable of sensing how their actions have affected others. He suggests that the life review could develop as a function of age. Research indicates that self-reflection develops gradually throughout childhood. However, children show behavioral changes following a near-death experience.

Neurological explanations, such as stress-induced memory responses or the brain's chaotic activation of autobiographical memory, are insufficient. What older children report feels like a precise structure, one that is clear, coherent, and embedded in a larger whole. Depth psychology refers to this as *integration*, while neuroscience speaks of *affective coherence*. Mystical traditions describe it as a space of knowing, where every act and every emotion leave a trace in the structure of all life: a moral echo.

Medically, it is clear: after cardiac arrest, the flow of oxygen ceases, neural networks collapse, and the electrical activity of the brain disappears. Under such conditions, consciousness should extinguish and not deepen. So how can such a coherent and clearly memorable experience arise in a state of clinical unconsciousness?

Hypotheses such as a "final discharge" of the dying brain explain this insufficiently. Many children report phases in which cognitive processing is impossible. Often, these accounts contain details that can later be verified. Therefore, near-death researchers such as cardiologist Pim van Lommel and psychiatrist Bruce Greyson propose a broader model: consciousness as resonance, not as a product.

In this view, the brain is not a generator but a filter. A receiver that opens in moments of near-death and allows access to a more expansive state of perception. Some researchers refer to this as the *nonlocal mind*, while others describe it as *field consciousness*. What unites them is the hypothesis that consciousness transcends the boundaries of the body.

Anna

The Second Threshold

Five years after the accident that had nearly taken her life, Anna stood at another turning point. She was twenty, studied psychology, and lived alone in a small city apartment. On the surface, she seemed structured, clear, and stable. However, something within her had begun to slow. It was not a crisis, a collapse, or a new trauma. It was more a gentle, almost imperceptible stilling.

"Something inside me grew quiet, too quiet," she said later.

She began to dream with a quiet insistence. The tunnel she had once passed through in her near-death experience reappeared. But this time, it was not filled with light. It was empty, not threatening, but open and waiting. She sensed that she was being called—not to die, but once more, to see.

During this time, Anna began to meditate again, at first out of restlessness, later from stillness. One evening, while sitting, something happened. She felt a loosening, as if her thoughts were slipping away; no image, no sound, only falling and being held. Then there was darkness and a point of light. It drew closer, or she did. She could not tell. And suddenly, everything was there: her childhood, the accident, the light, and the choice. That did not happen as a sequence or in parts but as a pattern.

She understood that her near-death experience had not been an isolated event. It was the beginning of an inner current, one that carried her, shaped her, and remembered her. Although no voice spoke in this second threshold, something was clear: She had not returned only to continue living. She had returned to see, and to remember.

"I didn't pass through the light in order to forget it," she said. "I passed through it to carry it with me."

After that experience, she began to write, not much, just sentences or fragments. As stated before, she began to accompany others, such as adolescents in crisis and people with threshold experiences, not with techniques but with presence. People said she "looked differently." And sometimes she would simply smile and nod.

"I look deeper," she said. "Because I once looked through the light. And because I recognize it when it appears in others."

Scientific Commentary

Some people who have had a near-death experience can mentally visualize and emotionally enter a near-death-like state during meditation. Such experiences could be referred to as "second thresholds" or post-integration phases. The focus is no longer on what happened, but on what remains. The near-death experience does not appear as a closed chapter but as part of a continuing inner journey. The "new identity" settles in, daily routines shift, and the sense of self quietly reorganizes itself.

Depth psychology speaks of archetypal reactivations, while mystical traditions describe it as an inner reunion. The memory returns, not because it is unresolved, but because it wants to deepen. The tunnel is not a relapse but a symbol, a movement, and a reconnection with a knowing that was never lost but had matured.

Many children report that their experience continues to have an effect years after the actual event. It does not end with the return to life but unfolds as an ongoing process that deepens over time. This "second threshold" indicates that near-death experiences do not merely mark a moment but serve as a catalyst: They open an inner path that matures by demanding to be lived in the present. Here, integration means bringing what was experienced into every breath, every encounter, and every day.

In states of very deep meditation, short or long vivid sequences of past memories, as well as structured life images that appear like a cinematic review may arise. These states offer a new perspective on a person's life or particular phases of it. These experiences tend to occur when the default mode network shifts into a different operating mode and a profound state of conscious openness takes place. The default mode network is associated with the autobiographical self. A life review during meditation can emerge through increased awareness, altered perception of time, the loosening of psychological defenses, and increased access to memory.

Jan

The Eye of Memory

Jan was twelve years old when he lost consciousness for several minutes during an epileptic seizure. His body lay motionless, his eyes rolled back, and his face became pale. A swift medical intervention stabilized his condition. The episode was deemed uncomplicated. And yet something had changed, not in the electroencephalogram (EEG), but in a layer that no instrument could measure.

In the days that followed, Jan began to speak of things no one could quite place. He talked about a grassy hill that smelled like summer and about a white house with blue shutters. He also described other things: a small stove that smelled of apples, a workshop with a glass lamp that shimmered in the sunlight, and an old man who sat across from him quietly smiling and wordless.

"He didn't speak," Jan said. "But I knew he was waiting for me."

His parents were puzzled. Had he dreamed it? Was it a story, a film, or a forgotten conversation?

Jan shook his head. He was certain: "I was there. Not yesterday. Not in this life. Before."

His parents did not give these comments much thought until the day, months later, when they walked through a mountain village in

Southern Italy during an unremarkable stop on a family trip. There were olive groves, dusty paths, and blooming poppies by the roadside. Suddenly Jan stopped. He stared at a tilted, faded, weathered house that seemed to be almost forgotten.

His voice was calm, almost factual: "I sat there. In the workshop. The man showed me how to draw with charcoal."

His mother held her breath. She vaguely recognized the house. It had once belonged to a quiet family acquaintance, an old man who used to draw, who spoke little, and who had no children. Jan could not have known.

And yet when a neighbor unlocked the door and they stepped inside, Jan walked straight through the room and pointed at a workbench.

"That's where the paper was. And the light was there."

In a dusty drawer, the mother found a faded folder. Inside were a few sketches and a single framed drawing which was delicately rendered. The drawing showed a large, striking, almost alive eye. Beneath it was written in barely legible handwriting: "See what does not fade." Jan looked at the drawing for a long time. He did not speak or cry. He simply stood there, as if he had arrived at a place recalled from memory.

Later, when his mother asked him how he knew this place, he answered:

"It wasn't a place. It was a part of me that came back."

From that day on, Jan began to draw scenes as well as specific fragments: an old hand holding charcoal, a slanted beam of light, and a face that never turned fully around. He said little but sometimes murmured:

"Some things don't come from now. They return when the time is right."

His drawings often featured eyes that were open, soft, and watching. Once he wrote beneath one of them:

"I think memory doesn't belong to time. It belongs to something deeper."

He never called it reincarnation. He never used big words, but something in his gaze had changed, as if a part of him was now seeing through what others simply looked at.

Scientific Commentary

Jan's story raises a question that reaches beyond any medical or psychological explanation: Can a person remember something they have never experienced? And if so—what, then, is memory? What if memory is not local?

In our familiar world, memory means something stored, retrievable, and laid down in the brain: a place, a time, or a feeling. Memory is coded in synapses, shaped by neural patterns, and chemically preserved in gray matter. However, some children who return from a near-death experience speak of something else. They recall things they never lived. Places they have never seen. People they have never met. Jan's statement, "Some things don't come from now. They return when the time is right," escapes the linear understanding of memory as a stored past. It points toward another layer where memory is non-retrospective and present, a memory without a place.

In Jan's case, there were none of the classical images of a near-death experience, such as a tunnel, a light, or otherworldly figures. And yet he brought back something from his brief unconsciousness, details marked by an unusual precision: a workshop filled with sunlight, a silent man, a drawing that had, in a sense, been waiting for him. These details might be dismissed as imagination, if not for the pivotal fact: Jan had never been there in his life. He had no prior knowledge and no biographical connection. And yet he recognized the place without hesitation, as if it had always been familiar to him.

Depth psychology provides preliminary interpretive frameworks for such phenomena. As mentioned earlier, Carl Jung's concept of the collective unconscious suggests the existence of shared archetypal patterns and memories that do not originate from individual biography. Transpersonal psychology refers to this as *transpersonal memory*, a form of knowledge that is not acquired but emerges from a deeper layer of consciousness.

The theory of morphic fields proposed by biologist Rupert Sheldrake also points in this direction: memory is not merely a neural retrieval but a resonance with non-material, timeless patterns. This means that memory resides in fields that are not bound to matter. Organisms, and possibly forms of consciousness, remember by tuning into existing patterns. From this point of view, even a relationship is not merely biographical but a resonance field that exists across time and space. When children speak of people of light, voices never heard, or places never visited, perhaps they are touched by a deeper reality that lies beyond thought and remains accessible to experience.

In quantum-inspired models, such as those of physicist Federico Faggin or psychologist Dean Radin, memory appears as a connection to a nonlocal field of information. From this latter point of view, Jan had not retrieved something from the past but had entangled himself with a place and an experience beyond time.

REFLECTIONS

Many children report experiences from phases in which no cognitive activity could be measured, and still they describe verifiable details of their surroundings. Anna's experience touches the limits of our current understanding of consciousness and memory. It shows us that memory is not merely a sequence of images but a space or a field in which time becomes depth and every encounter, every gesture, and every feeling is present all at once.

For Anna, it was not a film passing by her. It was an immediate experience. She said: "I was me. But I was also the others." There was no judgment or guilt, only connectedness. She felt how her actions resonated in others. It was empathy in its purest form: an entering into the lives of others until the boundary between I and you disappeared. Many reports confirm this pattern: the life review is not a chaotic flickering of the brain but a precise order. Medically, it is assumed that the brain quickly becomes inactive after cardiac arrest. Yet precisely in this state, when consciousness should actually vanish, these deep, structured experiences often emerge. How is that possible?

In this perspective, the life review is not a retrospective but a reconnection. It is less memory than insight: an entry into the interconnectedness of all life, in which nothing is lost. Anna's return was not only a return to life; it was a return to a deeper level of presence.

For Anna, the light did not end when she came back. It matured. It continued to accompany her in the quiet attentiveness to nature, to encounters, and to breath. She began not only to see the world but to listen to it.

This might be the true gift of such experiences: that the light, once encountered, does not fade. It remains as resonance, woven into the fabric of our lives. It teaches us that transformation does not happen in grand gestures but in the quiet moments when we are still enough to listen.

Maybe every child carries such a thread within them: a memory that does not begin in the brain but in the deeper rhythms of life. At times, we encounter a remembering that does not originate from our own life. A scent in the air. A certain fall of light. A glance through a stranger's window. Then, suddenly, we feel recognized by it. In such moments, it may not be we who remember, but the image that remembers us. A house that has been waiting. A chair that already knew us. A hand that once moved across paper, for someone not yet born. In other words, there may be places, faces, and images that wait for us, not because we have forgotten them, but because they belong to something that remembers us.

When Jan stood silently before the weathered door, that is precisely what happened. He did not understand. He felt. What unfolded within him was ancient and familiar. Open, without context. True, without proof. It needed no interpretation. It was simply there. Perhaps memory is not only something that we carry, but something that carries us. A field of remembering that does not belong to us, but in which we are held. A memory not bound to a single life, a body, or a self, but to a deeper resonance that moves through us and calls us back to what is waiting for us.

Perhaps we need to rethink memory, not as retrieval or a function of the brain but as a property of consciousness. Jan's experience suggests that some memories do not belong to us. They are preserved in the structure of the world and return when we are open enough to receive them. Seen this way, some memories are not echoes of the past but invitations from a timeless space. Instead of recollections, they return as quiet, insistent, and unprovable directions. They do not speak of what was, but of what has always been. And perhaps if we become still enough, if we stop trying to explain too quickly, we begin to see it: the window that waits. The eye that gazes. The handwriting of memory that does not meet us in the past, but precisely where the next step begins.

The Light That Carries Us All

The stories of children who have returned offer us keys to understanding, not to an afterlife but to a different way of being here. They suggest that consciousness is not produced by the brain, but expressed through it, without being confined to it. The children do not analyze these connections or interpret them. They remember. Their experiences are not overlaid with interpretative agendas. They are immediate, unfiltered, and free of philosophical and religious claims. The stories are less about definitively understanding than about lingering: listening to a question without answering it.

Therefore, this book does not end with a final point. It ends with a space where the voices of the children may resonate, because they have touched something within us that is older than words. Perhaps it was not the light on its own that moved us, but the echo of something familiar, something we recognized before we ever read it.

Hence, this final chapter opens a soft horizon. The children gave us insights into their silence, their resonance, their unadulterated closeness to the mystery of being. Their words are invitations to consider that consciousness is not just what we think but also what we feel and experience as part of a larger breath. The breath of the world that carried Adrian. Or in Amina's words: "something greater."

In those moments where the "ego" or "I" dissolves and something greater emerges, we encounter a truth beyond knowledge. These children remind us: We are not only those who ask. We are those who receive, carried within a breath that embraces us all.

What remains when we remove the word "end"? If we understood completeness not as an ending but as an arrival in a stillness that has always been present? The stories in this book remind us of a consciousness that was never entirely lost. A rhythm that carries us, even when we cannot define it.

The children speak of a vastness of consciousness that does not know boundaries, and yet it is so near that it touches our innermost being. In the resonance of their words, in the quiet depth of their experiences, we find a space where time is no longer a succession but a presence. A presence that is more real than the world we know, as Milos described it. Memory is not bound by the past but a living nearness. These children did not teach us to explain life. They taught us to feel it as a breath that moves us.

When we enter the space of these stories, we enter a place without fixed forms and a presence that is both familiar and beyond naming. Meera described it as an immediate, whole-being experience; Adelina calls it remembrance. Here, the "ego" or "I" loses its fixed form and becomes a part of a greater breath—flowing, embedded, and carried. Lev knew that one can be present even when one is not there.

As different as the individual stories are, clear common threads nonetheless emerge. In these experiences, consciousness is not perceived as something that arises within the body and perishes with it. Rather, it appears as something encompassing and relational: as a field in which perception, memory, and presence are interwoven.

In these moments, time loses its linear character. It is experienced as simultaneity rather than sequence. Past, present, and future do not unfold one after another, but are present within a single moment. Memory, in this context, is not bound to biography. It functions more like a form of closeness, something that is there even when we do not seek it.

Perhaps the greatest gift left by these stories is not the knowledge itself, but in what they teach us, that knowledge does not always need to be named. Not everything asks to be explained. Sometimes, it is enough to listen. To listen to what sounds between the words, lives between the thoughts, and waits between the breaths. And they invite us to listen to the questions within ourselves. They show us that true knowledge does not lie in knowing more but in listening more deeply. If you, dear reader, have come this far, you may also be a part of this dialogue. Perhaps you have recognized yourself in these stories, because they awakened an echo of something that has always been within you.

The children remind us of a quiet certainty that cannot be captured in words yet remains felt. Especially in twins, the relationship often does not end but changes its form, as is the case with Elias and Michael, Matthew and Elijah, as well as Oliver and Julian. A shared presence becomes a quiet, enduring bond. This presence calls us into remembrance that we are more than thoughts, more than bodies, more than consciousness in its limited sense. We are part of something greater—a light that carries us all.

Children After Returning – What Changes

The depth of these experiences becomes especially evident in what unfolds after the return. Many children come back from the brink of death without any outward signs of change—and yet, something is different. What happens to a child who returns from an inner threshold not necessarily caused by clinical death? A place that leaves no visible scar but leaves a mark.

Many of the children in this book speak differently afterward or fall silent, but in a new way. They observe more carefully. They dream more vividly. They pause before answering. Some need time to find words for something that cannot be easily explained. And often, without effort, they carry a kind of stillness. No withdrawal, no confusion, rather an expansion. Something within them has grown quieter. Others find forms of expression in words, drawings, dreams, or in a particular closeness to nature, animals, or to other people. These children do not appear traumatized. They appear touched. The behavioral change resembles a memory that continues to live on in everyday life. A certainty that does not need to be explained.

Magda speaks of an inner peace that settled in her chest. Catherine feels what others cannot see. Eli talks less about fear of death and more about returning. Marlon writes words that seem to have waited for him. Lilia and Liora speak of messages that were never said but fully known. None of these children seem disturbed. None speak of trauma. What remains is not pain but presence. A shift. A knowing.

This transformation is difficult to name. It is not a developmental stage, a coping strategy, or a "spiritual awakening" in the adult sense. It is more of a soft reorientation to something inward and subtle. Parents often speak of changes they cannot explain. Roya's mother calls it extraordinary sensitivity. Catherine's father notices a quiet patience. Many simply say: "Something is different." A look. A gesture. A breath that lingers a little longer.

In clinical psychology, this might be interpreted as post-traumatic adjustment. Nonetheless, often there is no trauma, only depth. What these children bring back is not a disorder but clarity. Gentle, unobtrusive, and yet unmistakable. Clarity that ranges from Noah and Timo, who saw the future, to Anna and Jan, who saw the past.

Perhaps these children are not different in the sense of being separate. Perhaps they are more open. And perhaps they carry within them something that we also once knew: the ability to know without explanation. To trust without proof. To receive without interpretation. To remember something that was never taught.

Liva and Mark even felt the need to recall a language unknown to them in life. Some children, like Emil, access this remembrance with the help of animals. Tiago and Toma sought it with the jaguar as their spiritual guide. Facing death, Matthias recognized the stillness of the world, and Dalia knew that someone was waiting for her.

They do not need interpretation, guidance, or reassurance. Maybe what they need is what they themselves offer: presence. A presence that listens without explaining, that trusts without demanding, that waits without pressing. Noémi understands what is essential without hearing. And Nia hears the sound of time. And perhaps they do not return to a different world. Perhaps they return to the same world, but they see it with different eyes. Mira and Elin perceive the world as harsher than before. Lenn, by contrast, entered a quiet peace with the world and his body.

Such experiences open a window onto a deeper layer of life, a layer that transcends the classical divisions between the self and the world, the body and the mind. They show us that the most powerful transformations are those that are lived and felt, even if they can't be described. Quietly. In the walk. In the gaze. In the breath. Where the light begins.

ACKNOWLEDGMENTS

This book would not exist without the courage and openness of the children and families whose experiences inspired these pages. The stories shared here, the depth of their encounters and the trust with which they were entrusted to us are the heart of this work.

We are deeply grateful to all those who dedicate their lives to exploring the mystery of consciousness, life, and death—researchers, physicians, hospice workers, therapists, and spiritual companions who approach these questions with curiosity, humility, and compassion.

Our sincere thanks also go to the team at Sacred Stories Publishing for their openness and thoughtful guidance throughout this project.

Finally, we thank one another for the shared journey of writing this book. What began as a collection of questions gradually became a deeper exploration of the human experience—and of the possibility that consciousness may extend beyond the boundaries we once believed to be final.

May this book encourage readers to listen more carefully: to the stories of children, to the quiet moments at the edge of life, and perhaps also to the deeper silence from which understanding begins.

Daniela & Ralph Klose

MEET THE AUTHORS

Daniela Klose is a medical writer and professional translator specializing in neuroscience, psychology, and consciousness-related topics. For more than ten years, she has worked as a ghostwriter and editor for researchers, clinicians, and academic institutions across Europe, supporting the development of scientific articles and nonfiction books.

Her work focuses on bridging complex scientific material with accessible language and emotional depth, with a particular interest in the intersection of neuropsychology and transpersonal psychology. Over the years, Daniela has contributed to a wide range of projects dealing with trauma, perception, altered states of consciousness, and the psychological dimensions of healing.

Besides her writing career, Daniela gained direct practical experience in retirement and nursing homes, where she encountered the subtle and often unspoken realities of transition and end-of-life awareness. These experiences continue to shape her approach to storytelling and research, especially when exploring the inner worlds of children and the ways consciousness expresses itself beyond conventional explanatory frameworks.

Ralph Klose is a retired neuropsychologist with more than thirty years of clinical and academic experience in the fields of cognition and consciousness research. Throughout his career, he worked with complex neurological and psychological conditions and developed a long-standing interest in the deeper mechanisms of perception, identity, and awareness.

Alongside his professional career, Ralph has spent decades supporting the development of academic and spiritual publications as a ghostwriter and research collaborator. His work includes contributions to books and articles across multiple disciplines, often helping authors translate scientific findings into precise, understandable, and intellectually rigorous narratives.

A defining part of Ralph's life has been his long-term volunteer work in hospice care, where he accompanied both adults and children in their final stages of life. These encounters offered him immediate insight into the psychological and existential dimensions of dying, as well as the remarkable accounts that sometimes arise at the threshold between life and death.

Now writing under his own name, Ralph combines scientific reflection with lived experience, exploring consciousness not only as a neurological phenomenon, but also as a profound human condition.

www.drklose.com

www.ingramcontent.com/pod-product-compliance
Lightning Source LLC
LaVergne TN
LVHW090400210926
848480LV00002B/402
* 9 7 8 1 9 5 8 9 2 1 9 1 3 *